# EVOLUTION:

## FACT,

## FRAUD, OR

## FAITH?

### DON BOYS, PH. D.

**FREEDOM PUBLICATIONS**
**P.O. BOX 981**
**LARGO, FL 34649**

Evolution: Fact, Fraud or Faith?
Copyright 1994, Donald G. Boys

All Rights Reserved

First Edition, 1994

Freedom Publications
P. O. Box 981
Largo, FL  34649

ISBN 1-879805-01-4

Library of Congress Catalog Card Number 93-72683


Printed in the United States of America.

All Scripture is from the Authorized King James Version.

# ABOUT THE AUTHOR

Don Boys was born in 1935 in West Virginia where he received his early training. He was educated at Moody Bible Institute, Tennessee Temple University, Immanuel College, Freedom University and earned his Ph.D. at Heritage Baptist University.

Boys has served in the Indiana House of Representatives, and while in the House, the media identified him as "the most conservative voice in the General Assembly." He authored legislation to require the public schools to teach creation on an equal basis with evolution; life in prison for rape; prison for sodomy; reversal of the E.R.A.; and other conservative measures.

Dr. Boys has authored nine other books: *Liberalism: A Rope of Sand; How to Feel Better, Look Younger, Live Longer; A Time to Laugh; Pilgrims, Puritans and Patriots; Is God a Right-Winger?; Christian Resistance: An Idea Whose Time Has Come--Again; Pioneers, Preachers and Politicians: The True Story of America; Boys' Big Book of Humor and AIDS: Silent Killer.*

Boys has written many feature articles that have appeared in major journals and from 1985 to 1993 wrote columns for the national newspaper, *USA Today*. He has been interviewed by major newspapers like the *Chicago Tribune, Los Angeles Times, Indianapolis Star,* the

*Sacramento Bee,* the *Chattanooga News-Free Press,* and many other papers and news magazines.

Dr. Boys has been a guest on major TV shows such as the "Sally Jessy Raphael Show" (twice), the "Jerry Springer Show" (twice), the "Morton Downey, Jr. Show," "Cross Fire," "NBC Nightly News," "CBS Morning News," "9 Broadcast Plaza," and many cable and syndicated shows. He has been a guest on most of the major radio shows in the major cities of the nation.

He and his wife currently live in Ringgold, GA.

# TABLE OF CONTENTS

# DEDICATION

This book is lovingly dedicated to my wife, Ellen Kate.

# FOREWORD

In his book, **Evolution: Fact, Fraud, or Faith?** Dr. Boys makes clear that as far as belief in the theory of evolution is concerned, evolutionists employ mostly a great deal of faith, then occasional fraud, and lastly very little, if any, real evidence. He pulls no punches, and goes at evolutionists and their belief system bare-fisted with a relentless barrage of facts exposing weaknesses and fallacies of the theory. In his anti-creationist book, evolutionist Niles Eldredge joins in the battle against creationists with the cry, "The gloves are off."[1] It is only fair, then, that in self-defense creationists should also strip off the gloves and go at the fraud of evolution with bare fists.

Soren Lovtrup, a well-known Swedish scientist and an unreserved evolutionist, nevertheless condemns the modern neo-Darwinian theory of evolution. This theory is being taught as fact in practically every major university in the United States. In his book, **Darwinism: The Refutation of a Myth**, he wrote, "I believe that one day the Darwinian myth will be ranked the greatest deceit in the history of science."[2]

Creation scientists go one step further and maintain that the theory of evolution, regardless of the mechanism suggested, is the greatest deceit in the history of science. Don Boys not only provides a wealth of evidence which supports special creation and contradicts evolution theory, but he also documents the fact that basically evolution is a religion and not science.

If you appreciate a hard-hitting attack on the religion of evolution dressed up in scientific clothes, you will enjoy this book in which Dr. Boys documents the fact that the emperor has no clothes at all.

Duane T. Gish, Ph.D., Vice-Pres.
Institute for Creation Research
Santee, California

# ACKNOWLEDGMENTS

No author can take full credit for writing a book since so many people are always involved from the concept to the finished product. This book has been a seven-year task that has taken me across the nation and border to border in research and writing.

During the writing of this book, my life has been totally uprooted with the death of my wife of thirty-three years, the marriage of my youngest child, the birth of two additional grandchildren, my marriage to a twice-widowed lady, and a move from Indianapolis to Georgia! No surprise that it took me seven years to finish the book.

I must express my appreciation to Dr. Duane Gish, Vice President of the Institute for Creation Research in California for reading the manuscript and making valuable suggestions, and for writing the Foreword.

Then Jennifer Ould did an excellent job proofing the manuscript, and making suggestions.

But the biggest thanks goes to my wife who read it over and over (and finding typos each time). She also made many suggested changes in content, and most of them were accepted. Thanks, sweetheart.

Of course, any problems, mistakes, or unconscious distortions are my own responsibility. I will sincerely appreciate any corrections from any person or group, and will make the necessary changes in the following edition.

# PREFACE

Evolution means "an unfolding process." An egg becomes a chicken, a bud becomes a rose, etc., but you never have a chicken without an egg nor a rose without a bud. Animals have "evolved" within a species, but you never see life come from death. And a species has never changed into another species, never!

Some have asked: Why another book on evolution and creation? And that's a reasonable question. I am responding to evolutionists like Stephen Gould who admitted that creationist-bashing was in order for our time. Well, Steve, I'm bashing back! His buddy, Niles Eldredge said that the gloves are off. That's acceptable to me. I'm willing to bare-knuckle it with any evolutionist. No quarter asked or given!

You will discover in these pages that brilliant people believe "dumb" things, and scientists who have hitched their wagons to evolution have hitched them to a falling star.

To change the metaphor, the evolutionary ship is going down, and hysterical scientists are frantically rearranging the deck chairs and polishing the brass handrails while the more astute ones are jumping overboard.

I do not mean to denigrate true science nor honest scientists for they have been largely responsible for many of the good things we enjoy in life; however, irresponsible scientists (with the zeal of Muslim fanatics) have perpetrated upon a gullible public the most gigantic fraud since Lenin preached the glories of his revolution to a hungry people.

The differences in evolutionists and scientific creationists are usually not in the facts, but in interpretating the facts, and I suggest that what you believe determines how those facts are interpreted. I'm a Bible-believing Christian, so I have made certain judgments, **and so has every evolutionist!** I accept the Bible as true, while evolutionists have accepted the philosophy of Darwin as true. (And it **is** a philosophy, not a science.)

In these pages, evolution is considered the gradualism of Darwin or molecules to monkey to man development by natural means. Since the 1950s Darwinism has been modified to mean the gradual development of the species through micromutations and natural selection interacting with the environment (neo-Darwinism). I will also deal with other off-shoots of Darwinism.

Scientific creationism will be considered to mean the origination of everything in the universe by divine fiat as recorded in Genesis 1 and 2.

If God did not do as He said he did, then He is unreliable, untrustworthy and unnecessary! If God did make us all, then we had better listen to Him in His Word and make it applicable in every facet of life.

You will discover that we creationists have far more than the Bible to support our position (although we don't need anything more). Biology, geology, paleontology, etc. all support special creation and refute Darwinism and neo-Darwinism.

# INTRODUCTION

Bernard Stonehouse, an evolutionist from Cambridge, has said of the "groundswell of opposition" to Darwinism, "...we should welcome it as an opportunity to re-examine our sacred cow more closely..."[1] In this book, I intend to examine the sacred cow **very** closely, and hopefully kill it and make some hamburger! Yes, it has been tried before, but **some** creationist must fire the final, fatal shot and kill the "beast."

Macroevolution (the change from a lower species to higher), as opposed to change **within** a species (known as microevolution), is not true. Darwinian gradualism is a fraud, a farce, a fake; and it has confused, contaminated, and crippled generations--especially children. I neither like nor respect those who harm people, especially weak, helpless, fragile children. Christ said that it would be better for a person who harms children to have a millstone tied around his neck and be thrown into the sea. There are so many evolutionists, we would run out of millstones, but I suppose we could use anvils.

There are many excellent books critical of evolution that are scholarly, calm and gracious, but I am convinced that we are in a battle for the minds and souls of men, so I have approached this subject like a war. And in this war I am not a conscientious objector. I consider evolutionists to be unrehabilitated criminals. I'm thoroughly mad and militant without being malicious. (Well, maybe a little.) Don't waste your time and money trying to convince me that, "Christians shouldn't be so militant." Such people obviously aren't familiar with the Old Testament prophets nor some of Christ's activities.

Of course, we Christians should always be "Christian." We should always be honest. We should be meek; however, we must not be **weak**! There is a difference, you know. I am told that we can "catch more flies with honey than with vinegar," but I'm not trying to catch flies. I'm trying to "kill skunks." I want to hinder, hamper and hurt the cause of evolution in any way possible as long as it is legal, moral and Christian. Some evolutionists will accuse me of being "unchristian," but that's like a skunk accusing a rabbit of having bad breath!

Evolutionists are perceived by the uninitiated as being calm, honest, fair and decent people--and a few are, as I will document in this book. But many evolutionists are vicious and vitriolic when it comes to creationists. I have also documented that **fact**. Those are the "skunks" that I will "kick around" in this book. Of course, some of the stench will be sprayed on me, but so be it.

One such "skunk," Thomas Huxley, was a hater of religion, and is known as "Darwin's bulldog." He wrote to Darwin offering to help his cause: "And as to the curs which will bark and yelp, you must recollect that some of your friends, at any rate, are endowed with an amount of combativeness which...may stand you in good stead. I am sharpening up my claws and beak in readiness."[2] I think Huxley had an attitude problem! When it came to Christians, he was as mean as a female pit bull with a bad case of PMS.

Another scientist with a problem, Garret Hardin of the California Institute of Technology, wrote that someone who doesn't honor Darwin "inevitably attract[s] the speculative psychiatric eye to himself."[3] So, I'm a candidate for the funny farm simply because I think Darwin was off base (or not even in the game). That's **very** scientific!

Sir Gavin de Beer, Director of the British Museum of Natural History, took a shot at sensible people when he wrote: "Only ignorance, neglect of truth, or prejudice could be the excuse for those who in the present state of knowledge without discovering new facts in the laboratory or in the field, seek to impugn the scientific evidence for evolution."[4] My intellectual opinion of de Beer is that he is showing his own ignorance!

As the foundations of evolution crack further and more evolutionists "jump ship," those remaining on board seem to get hysterical.  So they, in desperation, are doing, saying and writing some really "dumb" things (as this book will reveal).

Some very candid and honest **scientists** have characterized Darwinism as "speculation," "based on faith," similar to theories of "little green men," "wrong," "theory in crisis," "coming apart at the seams," "dead," "effectively dead," "very flimsy," "incoherent," and a "myth."

Well, with friends like that, no wonder the evolutionary ship is going down!

Author Norman Macbeth, not a creationist, admitted that the gradualism of Darwin is "sadly decayed." He added, "In brief, classical Darwinism is no longer considered valid by qualified biologists."[5]

Evolutionist  Sir  Arthur Keith confessed, "The only

alternative to some form of evolution is special creation, which is unthinkable."[6] "Unthinkable," why should any possible theory be unthinkable? What kind of science is that? (You will discover that it is **not** science but religion that prompts that prejudice.)

Well-known scientist Michael Denton admitted that "the number of biologists who have expressed some degree of disillusionment [with evolution] is **practically endless**."[7] (Emphasis added.)

Evolution--if not dead--is at least dying. Let's hope its demise is not lengthy, although we know the "faithful" are trying to resuscitate it. The only people who believe in the gradualism of Darwin are deceived children, fools, half-wits and college professors who get paid for teaching it. No educated, honest person believes the fairy tale of evolution any longer! The following pages will support that premise, so read on.

# CHAPTER ONE

# THE PRESENT DEBATE

Fanatics, like Julian Huxley and Carl Sagan, snort that evolution is "a fact" and only flat-earthers think otherwise. Such men, in my opinion, are not only fanatics but fools as well; however, their number is slowly decreasing as honest scientists jump from the good ship, **Darwin**. Honest scientists, after years of searching for answers, have become weary of the confusing, convoluted and contradictory claptrap that often passes as science. They have observed their colleagues rushing to protect Darwin rather than putting him to rigorous tests.

I salute those honest scientists who have broken with Darwin (you'll meet some in this book), and I give them a symbolic hug. I feel pity and pathos for those who are defending a castle in ruins (or to return to my earlier metaphor, who are clinging to the "**Darwin**" as it slips beneath the water). Those scientists, reading this book, will be bugged and slugged, maybe mugged.

As to the inane, inaccurate and insipid suggestion that evolution is a **fact**, even Michael Denton, international scientist replied, "Now of course such claims are simply nonsense. For Darwin's model of evolution is still very much a theory and still very much in doubt...."[1] But insecure scientists seldom admit that fact except to other insecure scientists.

## THE SCIENTIFIC METHOD

Evolution, rather than being a fact, is a faith, since it can't meet the test of science. The Big Lie told by evolutionists in every debate between creationists and evolutionists is that creation science is religion while evolution is science. Of course, **neither** can be proved scientifically since neither can be tested by the scientific method.

The scientific method is "an orderly method used by scientists to solve problems, in which a recognized problem is subjected to thorough investigation, and the resulting facts and observations are analyzed, formulated in a hypothesis; and subjected to verification by means of experiments and further observation."[2]

Well, if words still mean what Webster and others say they mean, then **no theory** of origins can be proved scientifically because the essence of the scientific method is founded on observation and experimentation. One does not have to be a scientist to know that the origin of the universe can not be observed or put to experiment since it is not happening in the present. Even a Harvard professor can understand that.

Another important part of the scientific method is that any scientific theory must be capable of falsification. That means it must be possible to devise an experiment, the failure of which, would disprove the theory.[3] It must be testable to be scientific.

All right, we admit that creation was not observed

(by humans), can not be put to an experiment and is nonfalsifiable; however, the **same** is true of evolution! Macroevolution has never been observed, can not be put to experiment and is nonfalsifiable; consequently, evolution is not scientific. It is religion!

One of the best known evolutionists, G. G. Simpson stated: "It is inherent in any definition of science that statements that cannot be checked by observation are not about anything...or at the very least, they are not science."[4]

Author Norman Macbeth admits that "Darwinism is not science,"[5] while L. C. Birch and P. R. Ehrlich agree that evolution "is 'outside of empirical science' but not necessarily false. No one can think of ways in which to test it."[6]

Scientists are presumed to be people who have looked at **all** available evidence, and have made sane, sensible and scholarly calculations in various fields, consequently they can be trusted; however, some scientists who have not followed proper scientific procedures have made fools of themselves.

Lord Kelvin ridiculed x-rays as an elaborate hoax!

Ptolemy affirmed that there were 1,058 stars, but Kepler counted only 1,005! (Of course, Jeremiah told us that "...the host of the heaven cannot be numbered." How did Jeremiah know that scientific fact thousands of years before the telescope was invented?)

Until a hundred years ago, physicians thought they could safely examine a pregnant woman without washing their hands even after handling a corpse!

Experts told Americans in the early 1900s that humans must be careful of riding in automobiles, because the human body could not live at speeds past 35 miles per hour!

The above mistakes and hundreds of others are proof that scientists don't always follow the scientific method. In fact, evolutionists **butcher** the method in their labs while they **boost** it with their lips! There is no real contest between science and Scripture, only between Professional Speculators and creation scientists. When evolutionists are backed into a corner by pushy creationists, they usually stick to their guns-- even though they aren't loaded! They are usually very hesitant to admit their mistakes, then they whine when a creation scientist points out their errors. (Evolutionists whine much about being misquoted, but all we do is point out their mistakes, contradictions and foolishness, all in context.)

## RECAPITULATION THEORY

For example, Ernst Haeckel's "recapitulation theory" has been an embarrassment to honest scientists for many years. Haeckel formed his "biogenetic law," asserting that the human embryo re-enacts the various stages of past evolution.

Haeckel and others taught that an unborn baby went through a gill stage, tail stage, protozoan stage, etc. to the human stage. We were supposed to see our ancestry in the fish, frog, reptile, etc. stages in each embryo.

One example will suffice. Each embryo is said to possess "gill slits" during its development thereby proving that man was at one time a fish! But they are grooves not gill slits, and Dr. Duane Gish asks, "If they are neither gills or slits, how then can they be called 'gill slits'?"[7] Good question! Gish added, "These structures actually develop into various glands, the lower jaw, and structures in the inner ear."[8]

Haeckel was a zealot, a fanatic who believed his

cockamamie theory so strongly that he faked evidence to support it! He was tried and found guilty in a university court in Jena, but did not lose his position as professor! His "theory" has been discredited for over fifty years, but there are still a few poorly trained (or very naive) scientists who use the recapitulation theory to support evolution! The theory has been watered and manured for over a hundred years by incompetent and inept scientists who have lost the ability to blush.

Singer ripped Haeckel to shreds in his **A History of Biology**: "For a generation and more he purveyed to the semi-educated public a system of the crudest philosophy--if a mass of contradictions can be called by that name. He founded something that wore the habiliments of a religion, of which he was at once the high priest and the congregation."[9]

The harm done to true science by Haeckel's scatterbrained theory is enormous. Walter Bock of the Department of Biological Sciences of Columbia University admitted: "...the biogenetic law has become so deeply rooted in biological thought that it cannot be weeded out in spite of its having been demonstrated to be wrong by numerous subsequent scholars."[10] Yes, there is a price to pay when trained professionals jump to desired conclusions without supporting evidence.

Evolutionist Sir Gavin de Beer commented on the harm done by Haeckel: "Seldom has an assertion like that of Haeckel's 'Theory of Recapitulation,' facile, tidy, and plausible, widely accepted without critical examination, done so much harm to science."[11] Then, in his article, "Embryology and Taxonomy," he concluded that the theory had been discredited and generally abandoned.

Even the very liberal **Encyclopaedia Britannica** asserts that the theory is "in error,"[12] while Danson says it is "intellectually barren."[13]

Dr. W. R. Thompson, who wrote the Introduction to Darwin's 1956 **Origin** said, "The 'biogenetic law' as proof of evolution is valueless."[14] That statement should be final and fatal to Haeckel's pseudo-science, but alas, it seems not so.

Keith S. Thomson struck another blow to Haeckel in an **American Scientist** article when he opined, "Surely the biogenic law is dead as a doornail."[15] Well, it may be dead, but some evolutionists still teach it, such as Richard Carrington who wrote, "This principle that the individual embryo relives the story of the race, is known as the law of recapitulation. It is perhaps **the most conclusive proof** we have of the truth of evolution."[16]

Can you believe that a scientist would say that this almost universally rejected and scorned theory is the "most conclusive proof of evolution"? Well, you can see how much trouble evolution is in.

You wouldn't expect the "smartest man in the world" to fall into the recapitulation trap but, alas he did. **Parade Magazine** reports that their readers have chosen Dr. Carl Sagan, astronomer at Cornell University, as the smartest man in the world! And he fell or jumped into the quicksand of the "biogenetic law."

Sagan and co-author Ann Druyan wrote concerning the developing human embryo: "By the third week...it looks a little like a segmented worm. By the end of the fourth week...it's recognizable as a vertebrate, its tube-shaped heart is beginning to beat, something like the gill arches of a fish or an amphibian have become conspicuous, and there is a pronounced tail. It looks something like a newt or a tadpole....By the sixth week...the eyes are still on the side of the head, as in most animals, and the reptilian face has connected slits where the mouth and nose eventually will be....By the end of the eighth week the face resembles a primate's but is still not quite human."[17]

Can you believe that? Carl Sagan promoting Haeckel's recapitulation theory is like physicians endorsing practices of jungle witch doctors. I wouldn't trust these "scientists" to walk my dog!"

But surely Darwin tested the theory and rejected it! No, he thought Haeckel's theory was "evidence second to none" to support evolution.[18]

Many scientists rallied to this theory with gusto, and it was dominant in biology for many years, but alas, the science of genetics has  revealed its absurdity. It is time to "pink slip" those pathetic evolutionists who **still teach** Haeckel's discredited, disproved and disreputable theory.

## USELESS ORGANS

Another major error made by evolutionists is that of allegedly useless organs in man that are  supposed to be withered memorials of man's past evolution.  As late as the 1960s, evolutionary textbooks listed over 200 "useless" organs, but later information has proved that almost all "vestigial organs" have some function during our lifetime. The tonsils, appendix and thymus gland are known to protect us against infections especially during younger years.

Irreparable harm has resulted from this teaching as hundreds  of thousands of tonsils, adenoids, appendixes, etc.

have been removed from children and adults in the past hundred years. How much research has **not** been done on various "useless" organs because scientists taught **other** scientists that many organs and glands were useless.

Only four or five organs are now believed, by some evolutionists, to be unnecessary and these are questionable. And of course, just because an organ may not be **necessary** does not mean that it is useless.

Scadding, a zoologist, asserted, "The 'vestigial organ' argument uses as a premise the assertion that the organ in question has no function. There is no way, however, in which this negative assertion can be arrived at scientifically....I conclude that 'vestigial organs' provide no special evidence for the theory of evolution."[19]

If evolutionists had followed their much touted scientific method, they would not have made fools of themselves so many times. An honest scientist will keep an open mind so that he will not "jump to conclusions," (often **wrong** conclusions), until he has made **many** observations with the same results. He must be very careful to base any conclusion on what he has **actually** seen, rather than what he **wanted** to see. And he must always be willing to change his mind. To the credit of most evolutionists, they have done that regarding the recapitulation theory and vestigial organs.

I suggest that when honest, competent people look at Darwinian evolution (molecules to man) and compare it to creationism, creationism will always be preferable. We must insist that scientists do that which makes them as uncomfortable as a dog in hot ashes: they must look at the evidence and evaluate it without any presumption about the truth of the theory they are testing.[20]

Even Thomas Huxley, "Darwin's Bulldog," acknowledged that " '...creation,' in the ordinary sense of the word is perfectly conceivable. I find no difficulty in conceiving that, at some former period, this universe was not in existence, and that it made its appearance in six days (or instantaneously, if that is preferred), in consequence of the volition of some preexisting Being."[21]

Well, that's some admission from an evolutionist and religion hater! Maybe the "bulldog" lost some of his teeth during his battle with creationists.

Evolutionist R. D. Alexander, Professor of Zoology at the University of Michigan, stated: "No teacher should be dismayed at efforts to present creation as an alternative to evolution in biology courses; indeed, at this moment creation is the only alterative to evolution. Not only is this worth mentioning, but a comparison of the two alternatives can be an excellent exercise in logic and reason. Our primary goal as educators should be to teach students to think and such a comparison, particularly because it concerns an issue in which many have special interests or are even emotionally involved, may accomplish that purpose better than most others."[22] Well, well, another honest evolutionist! Terrific!

Darwin wanted people to view both sides of this issue, writing, "I look with confidence to the future--to young and rising naturalists, who will be able to view both sides of the question with impartiality...."[23] It is happening, but don't worry about being caught in a stampede of scientists eager to look at **both sides**. Our side scares them to death! And, if they don't look, they can't be convinced!

I am asking and expecting some evolutionists to display objectivity as they read this book, but when it comes to evolution, objectivity is as scarce as white dinosaurs in Kentucky!

While I was a member of the Indiana House of Representatives, I introduced a bill requiring public schools to teach evolution and creation on an equal basis. I had thousands of letters sent to fellow-legislators and Governor Bowen's office. He supported my bill and it passed overwhelming, only to be killed by the liberal president of the Senate.

After it passed the House, I contacted the head of the ACLU to get their support. (I hope I don't lose my conservative credentials for that.) I told their leader that the ACLU had a reputation for fairness, equity, etc. (I almost gagged on my words) and would they support my bill since I was only requiring fairness? They decided not to do so! Surprise, surprise, surprise!

It is interesting to note that the ACLU funded the Scopes Trial in 1925, and their pitch was that in all the schools in Tennessee, only **one** theory of origins could be legally taught. That wasn't fair, they whined. Well, what has happened since then? Very simple, they hold the monopoly in the schools today, and they don't want to let go. In other words, they are not interested in fairness. They are hypocrites, prejudicial hypocrites, in my opinion.

Scientist D. M. S. Watson displayed his prejudice when he wrote that evolution is "a theory universally accepted not because it can be proved by logically, coherent evidence to be true, but because the only alternative, special creation, is clearly incredible!"[24]

And, I suppose the fairy tale of evolution is credible! Obviously Watson is not a deep thinker and should be considered suspect even when quoting the multiplication tables.

Soren Lovtrup, Professor of Zoo-physiology at the University of Umea in Sweden, made a cogent assessment of Darwinism when he wrote: "I suppose that nobody will deny that it is a great misfortune if an entire branch of science becomes addicted to a false theory. But this is what has happened in biology: for a long time now people discuss evolutionary problems in a peculiar 'Darwinian' vocabulary-- 'adaptation,' 'selection pressure,' 'natural selection,' etc.-- thereby believing that they contribute to the explanation of natural events. They do not, and the sooner this is discovered, the sooner we shall be able to make real progress in our understanding of evolution.

"I believe that one day the **Darwinian myth** will be ranked the **greatest deceit** in the history of science. When this happens many people will pose the question: How did this ever happen?..."[25] (Emphasis added.) Well, much of the blame must fall at the feet of clergymen who were intimidated by incompetent evolutionists following the publication of Darwin's book.

## DARWIN DOUBTED

But we must remember that some of the greatest scientists in the world have rejected the gradualism of Darwin and have been courageous enough to do so publicly. One such man, considered the greatest scientist in France, is Pierre Grasse, who wrote **The Evolution of Living Organisms** to "launch a frontal assault on all forms of Darwinism."[26] His work is devastating to Darwinism.

Scientist B. Leith admits that Darwinism is "under attack." He wrote, "The theory of life that undermined nineteenth-century religion has virtually become a religion itself and in its turn is being threatened by fresh ideas....In the past ten years has emerged a new breed of biologists who are considered scientifically respectable, but who have their doubts about Darwinism."[27] Of course, **every** honest scientist has doubts about it, while others, like Stanley below, even admit that gradual change has **not** taken place!

Paleontologist Stephen Stanley of Johns Hopkins University wrote, "The known fossil record fails to document **a single example** of phyletic evolution accomplishing a major morphologic transition and hence offers no evidence that the gradualistic model can be valid."[28] (Emphasis added.) Note that he said, "no evidence" for gradualism. Just what we creationists have been saying for many years.

Surely all evolutionary scientists are not dishonest or incompetent or even bigots. I agree. Some have simply not looked at the evidence, while others have been overwhelmed with doubts but have decided to keep quiet, hoping those doubts will go away. Most evolutionists, I believe, assume that while evolution cannot be supported by the evidence **in their own field** (of geology, paleontology, etc.), it **must** be provable by other specialists in **their** fields. Afraid not.

Paul Lemoine, one of the most prestigious scientists in the world seemed to agree with the above when he wrote, "The theories of evolution, with which our studious youth have been deceived, constitute actually a dogma that all the world continues to teach: but each, in his specialty, the zoologist or the botanist, ascertains that none of the explanations furnished is adequate...."[29]

It is finally a time for truth: Evolution was wrong, is wrong, and will always be wrong. It is not a fact, but a fraud and a faith. Hsu, a well-known geologist at the Geological Institute E.T.H. in Zurich, says that Darwin didn't understand geology, and bravely concludes: "We have had enough of the Darwinian fallacy. It is time that we cry: 'The emperor has no clothes.'"[30] I agree. Darwin is naked as a jaybird!

All right, evolution is in serious trouble (breaks my heart). While some scientists have become creationists, others simply refuse to take a position on the subject, knowing it is not politically correct to confess that Darwin is not resplendent in a magnificent suit of clothes. Those scientists keep quiet, hoping they will not be discovered as anti-evolutionists because their bigoted colleagues will then relegate them to the lunatic fringe. I call those people, "closet creationists," and I hope to force some of them out of the closet. I at least want to open the door for them.

Other scientists, like Carl Sagan, are asinine, arrogant and audacious as they con a gullible public into assuming that all **thinking** people accept the humbuggery of evolution as a fact.

Still other scientists will tell you that, while evolution isn't a fact or a perfect theory, it is "better than the others." But so what! That is not particularly praiseworthy if the others are no good. That's like being exultant in outrunning a cripple in a foot race![31]

With the evidence already presented, plus the additional facts in the following pages, it is astounding to me that **any** sane, sensible, scholarly person would support macroevolution, but evolutionists seem to think that a **bad** theory is better than no theory (especially creationism). A.

Koestler was correct, "they are unable or unwilling to realize that the citadel they are defending lies in ruins."[32]

What a pity! Qualified scientists, in various fields, could accomplish so much more if they were not preoccupied with defending the indefensible, the fraud of evolution and accept the fact, supported by science and Scripture, that God created everything.

# CHAPTER TWO

# DID GOD CREATE EVERYTHING?

Until about 1850 almost everyone believed that God created the universe, our world, and all that is therein in six literal days, as recorded in Genesis one. Chapter 1 begins, "In the beginning God created the heaven and the earth." Many people assume that only ignorant, "religious" people believe in creation; however, that is not true. The best known and most respected scientists of the past were creationists, and many contemporary scientists are as well.

Sir Ernst Chain, a Nobel Prize winner for his work with penicillin, rejected evolution, as did Louis Vialleton, who was Professor of Zoology, Anatomy and Comparative Physiology at Montpelier University in France.

Professor Louis Bounoure, former President of the Biological Society of Strasbourg and Director of the Zoological Museum wrote: "Evolution is a fairy tale for grownups. This theory has helped nothing in the progress of science. It is useless."[1]

Then there was Dr. W. R. Thompson, an anti-evolutionist, who had such international status, that he was asked to write the introduction to the 1959 edition of Darwin's **Origin of Species**, even though he was very critical-- though gently--of the book! He "gently" demolished Darwin!

Sir Ambrose Fleming, physicist and inventor of the thermionic valve (which made high-quality radio broadcasting possible), was a vocal opponent of evolution.

Albert Fleishman, Professor of Zoology and Comparative Anatomy at Erlangen University, Germany said: "The Darwinian theory of descent has not a single fact to confirm it in the realm of nature. It is not the result of scientific research but purely the product of imagination."[2] How about that from a world renown scientist?

Dr. A. Thompson came out for creation in his book dealing with biology, zoology and genetics as they relate to creation. He wrote: "Rather than supporting evolution, the breaks in the known fossil record support the creation of major groups with the possibility of some limited variation within each group."[3]

Dr. Austin Clark was curator of paleontology at the Smithsonian Institution, and forcefully said, "Thus so far as concerns the major groups of animals, the creationists seem to have the better of the argument."[4]

Delving into the past, I could provide a huge number of great scientists who were creationists, but a few of the best known will suffice: Michael Faraday, Lord Kelvin, Joseph Lister, Louis Pasteur, Isaac Newton, Johannes Kepler, Louis Agassiz, Sir William Ramsay, Lord Francis Bacon, Samuel Morse, Georges Cuvier, etc. Obviously, the great men of science were creationists.

One of creationism's opponents is H. S. Lipson, an agnostic physicist and Fellow of the Royal Society. He rejects creation, yet thinks scientific evidence is supportive of creation rather than evolution! He wrote, "I think, however, that we must go further than this and admit that the only acceptable explanation is creation. I know that this is anathema to physicists, as indeed it is to me, but we must not reject a theory that we do not like if the experimental evidence supports it."[5] Wow! With enemies like that....

So, creation is not merely "religious." It is also science, meaning "knowledge," or "to know."

Now, if men would only take God at His Word, that would settle the issue, but sinful man has been trying to remove God since the beginning. You may remember that Satan asked Eve, "Hath God said?" Men are still asking that question, and lost men are answering, "No, He hasn't." The bottom line of evolution is an attack upon the reliability of the Word of God.

After all, if there were no creation, then there was no Creator! By removing God, man "solved" his biggest problem--personal responsibility to a holy God. With God gone, men could now live as they pleased, forget God (who didn't exist, according to them), and have no fear of a future judgment. So modern day evolutionists must never even consider the **possibility** of a personal God who created everything. They fanatically cling to the **Darwin** although everyone knows the ship is sinking.

What surprises me is not that so many crew members have jumped overboard, but that **anyone** is still on board! The mast has been shot off, sails are ripped, the ship lists to starboard and is shipping water. Of course, the ship was full of holes when Charles Darwin launched it in 1859!

Evolutionists pretend to be willing to consider any and all possibilities, put them to the test, and make a rational judgment; however, when it comes to special creation, they refuse to do so, because they must never admit that God exists. This proves that what you believe determines what you do with the available evidence. Scientists present themselves as objective seekers of truth; however, those who are atheists can not consider the possibility that God created anything for then they would lose their atheist credentials that admit them to the better "clubs," and expose themselves to professional ridicule. Thus, they lose their "objectivity."

Let me repeat my quote from Professor D. M. S. Watson who admitted as much when he wrote, "...the Theory of Evolution itself [is] a theory universally accepted not because it can be proved by logically coherent evidence to be true, but because the only alternative is special creation, which is clearly incredible."[6]

But then there is Nobel Prize winner, Dr. Robert Millikan who declared, "To me it is unthinkable that a real atheist could be a scientist."[7] So two highly trained scientists each think the other is out in left field. Why? It is their philosophy. What you believe determines what you do with the available evidence. I choose to believe in God, and while that doesn't change the facts, it does change how I interpret the facts. I believe God began it all.

## GOD SPOKE

Genesis clearly tells us that God spoke and the universe came into existence. Out of the darkness, God commanded, "Let there be light," and there was light. Skeptics ask, "When? Where? How?" How could light preceed the sun and stars, but that is no problem to God who is the Light of the World.

Uninformed infidels love to assert that there are two

different accounts of creation in Genesis 1 and 2, but they are "reaching" in desperation for some weakness in the Bible truth of origins. The second chapter of Genesis is simply a different perspective on general creation with emphasis on man's creation and his environment. Genesis 2 is a recapitulation of the events of creation and is not told in a chronological order. No big deal, but it displays the ignorance and folly of unconverted men who pervert Bible truth to assuage their conscience and dispel recurring thoughts of coming judgment for their unbelief.

At the Word of God, the Earth was thrust into space, the thick folds of blackness receded and flashes of brilliance filled the atmosphere at the speed of light. And God (who is Light) saw that it was good. So ended the first day of this world.

On the second day, God created the firmament--the expansion of the heavens. At this time, the earth was "without form," empty, without grass, trees, animals and people. In fact, it was completely covered with water. God, being aware of future evolutionary attacks upon His creation, added, "And God made the firmament" (verse 7).

On this second day of creation, God separated the waters of the oceans and the waters above. Verse 7 of chapter one says: "And God made the firmament, and divided the waters which were under the firmament from the waters which were above the firmament: and it was so." Most informed Christians believe that this verse tells of a great water vapor blanket that encompassed the earth (similar to the one around Venus), filtering out dangerous rays from the sun while permitting the sun's light and heat to bounce off the globe. This canopy of water produced a hot house effect--moderate weather all over the earth. Many think the water canopy's filtering of the sun rays was one reason early people lived hundreds of years. We are reminded that men lived fewer years immediately after the Flood of Noah, a result of the burst canopy.

It is now day three and the Lord divided the land and the water making the world a habitable place. The seas rushed into their nesting places revealing the earth, without trees, flowers or animals. God added color by covering the earth with grass, herbs and fruit trees. A banquet for the human race was being spread before the first guest arrived!

Genesis clearly teaches that the grass, herbs and trees each carried in themselves the ability to perpetuate "after his kind." The apple, pear and peach trees that exist today have descended from those fruit trees created by God! Those trees were also created with age. They were mature trees with fruit hanging on the limbs.

Can you imagine a factory that takes dirt, adds some water and sunshine and produces delicious fruits and nuts? All this has been going on since the third day of creation without any noise and without any machinery breaking down! And without any pollution! In addition to the fruit, God arranged for the leaves and plants to produce the oxygen we need and to consume the poison we discharge through breathing!

On day four of this world, God created the sun, moon and the stars for a reason: to give light, to separate day and night, for navigation, etc. Evolutionists believe every-thing happened by chance, while Christians believe there was a purpose to everything. God had a purpose for man's good. The moon, sun, stars, etc. were for light, heat, navigation, seasons, weather. The ancient Babylonians believed that the heavens controlled mankind, but Christians believe the heavens are for our benefit, not control.

Many pagans have worshipped the sun for centuries, considering it the source of life and health; however, millions of us have found that life, health, happiness, security, peace and joy are found in the Son, not the sun. Many unbelieving evolutionists point out that it takes vast periods of time for the light of distant stars to reach the earth, ipso facto, God could not have created the universe, earth, and man only about ten thousand years ago. Evolutionists tell us that the light from the closest star could not have arrived on Earth even now! However, God created adult trees, bushes with berries, and Adam with maturity. So evidently He created stars with light--mature stars with light already en route to the Earth!

In Genesis 1:16 we read, "...he made the stars also." And in verse 17, He said, "And God set them in the firmament of the heaven to give light upon the earth." Why would God say that if the light would not appear for billions of years in the future? Obviously, He created the stars **and** their light simultaneously.

Today is Friday, the fifth day in the life of the world, a world wrapped in brilliant green, one of God's favorite colors. Roses, buttercups, azaleas, spruce, pine, walnut and hundreds of other bushes, flowers and trees covered the earth. There was water, but not a fin flashed in the sunlight and not a wing soared overhead. But wait, now there is a splash in the four rivers of Eden, and a whir in the air. The air and seas are filled with millions of creatures, all created by God.

Verse 21 tells us that God "created great whales," not immature, baby whales. Here, again, God has created them with age as He did the trees, stars, etc. He also set definite parameters that they reproduce "after their kind." And day five ended.

The sixth day began with the dawning of a beautiful morning. The seas were full of flashing fin and strange creatures, and the sky was crowded with feathered creatures, but the earth had not one inhabitant. There was no one to pluck a flower, skip a rock across the river or enjoy the sunset, but God had a plan.

God created insects and animals, then forever required them to reproduce "after their kind"; that is horses would mate with horses, dogs with dogs, cats with cats, etc., and a cat would always give birth to a cat and **never** a dog. Then He made a man from the dust of the earth. God molded a perfect man, made in His image, but he lay lifeless on the grass until God breathed into him the breath of life. Then he stood erect.

The first man upon the face of the Earth looked toward God with gratitude. His eyes were full of heavenly light, and peace and serenity enveloped him. He watched the harmless animals play, the birds flit back and forth, and the graceful fish cut the waters. And all creation knew this man was in charge.

As Adam looked around, he discovered that he was alone, and God said that it was not good for man to be alone. So He put Adam to sleep, removed a rib from his side, and made him a wife. Evolutionists laugh at that fact, yet they tell us man has slowly evolved from a one cell amoeba to an astronaut! And we aren't expected to laugh.

When Adam saw Eve, he smiled again, grabbed her hand, and probably walked along the banks of the Euphrates River toward a bower of wild rose and honeysuckle. They stopped and listened to the call of the whippoorwill from the aromatic flower garden, and as the sun sank beneath the horizon like a giant ball of fire, the "evening and the morning" were the sixth day.

## SPECULATION

Bible critics, always trying to denigrate the Bible, tell us that Adam and Eve were mythical characters who only lived in the minds of simple, confused, uneducated people. Of course, they must take that position because their whole foundation would crumble if Adam and Eve actually lived. Well, the Apostle Paul believed in Adam and Eve for he said in I Corinthians 11:8, "For the man is not of the woman; but the woman of the man."

Then Paul told Timothy in I Timothy 2:13, "For Adam was first formed, then Eve." So, the Apostle Paul believed that God created Adam and Eve. Christ also endorsed Adam when He said in Luke 3:38, "....Adam, which was the son of God."

Soon after their creation, Adam and Eve put their innate knowledge to work. Adam's son, Cain built a city according to Genesis 4:17, and soon Adam and his descendants were working with minerals, mining them, smelting them and creating all kinds of tools, weapons and musical instruments.[8]

Evolutionists talk endlessly of ancient man's pursuit of animals for food, moving into caves when the ice age

encroached upon them, and then learning the use of fire.
They used a long bone to crack open a skull, and discovered
a "tool." They began to construct tools and weapons, then
emerged from their caves standing upright. Well, almost
upright. Later, they kept animals and domesticated sheep and
dogs, and then planted some crops. Man was beginning to
settle down, so the story goes, but that is pure speculation.

He dragged his belongings behind him until one day
someone put crude wheels on the contraption, and we had
the prototype of the first Ford. Suburbs here we come!
About this time he learned to communicate with his "peers"--
more than pointing and grunting. He talks! We have the
beginning of a politician. That is not science; it is not even
good speculation. It's also not history nor in the Bible.

Whenever you find man in history, you find intelligence, art, music, writing, speech, etc., and not a group of
grunting brutes dragging women by their hair into a dirty
cave. Dr. Cyril Smith of the Massachusetts Institute of Technology supported the premise of early man's culture and
learning when he wrote, "Figurines were certainly being fired
by 9000 B. C. in the Middle East....Ceramics was well on its
way to becoming the noblest of the pyro-technical arts once
the pot was seen to be not only useful but pleasing to the eye
and potters found that both the beauty and the utility of their
product could be increased by firing at higher temperatures
and by admixing various minerals to give rich color and
textural effects....Even today we understand only 'in principle' the whole range of physical and chemical properties
and their interrelationships that were effectively used in
making ceramics."[9] Sounds like high culture to me.

Sheep were domesticated by about 9000 B. C., long

before the dog or the goat.[10] You may remember that Abel, son of Adam, was a keeper of sheep.

Dr. Henry Morris documents that copper beads were found in Iraq dating from the middle of the ninth millennium B.C.,[11] and metal objects were found in Turkey dating before 7000 B.C.[12]

Written language goes back to at least 4000 B.C. So men, at their first appearance in history, are educated, cultured and not in caves! God's production of man was done right. Too bad we went wrong. Man has not been climbing up, but falling down. Evolutionists, as usual, have it wrong.

After a fantastic job of creation, God rested on the seventh day--a picture of the command He gave to the Jews to observe as their day of rest. That command was given only to the Jews and was never repeated in the New Testament, and the breaking of it is never listed along with other sins in the New Testament. As New Testament Christians, we voluntarily observe Sunday in remembrance of our Savior's physical resurrection from the dead. Of course, Adam and Eve did not receive the Sabbath commandment from God. (That command was given in Exodus 20:8, 11.)

Evolutionists tell us that Adam and Eve did not actually live, but were mythical characters who represent all mankind as man climbed out of the slime one million years ago. But Christians know that if they were not historical people then the Fall of man did not take place, men did not become sinners, and consequently a Savior was not necessary!

Now can you understand why unbelievers will not accept the Genesis account of creation? Their wicked hearts must exclude God from the universe making Adam and

Eve mythical characters, eradicating sin from the earth and making Christ only a good man with some nice teachings.

If Adam and Eve were not actual people, then Jesus lied and is not worthy of worship! In Matthew 17:4, Jesus was discussing marriage and divorce with the Pharisees when He asked: "...Have ye not read [citing Genesis 1:27], that he who made them from the beginning made them male and female...." Here Christ makes it clear that there **was** a beginning, that the first couple was "made," and that they were "male and female." That is hardly Darwin's evolution. Jesus also said in John 5:46-47: "For had ye believed Moses, ye would have believed me...but if ye believe not his writings, how shall ye believe my words?" If men won't accept Moses in Genesis, they won't accept Jesus in John.

There is absolutely no reason to believe that the first people were knuckle-walking brutes who sat around a flickering fire in the mouth of a dirty cave grunting to each other. Adam and Eve were created by God as Genesis reveals, and all the wild speculations of unbelieving evolutionists will not change the fact that "In the beginning God created the heaven and the earth" (Genesis 1:1).

But didn't ancient men live in caves? Men have lived in caves during all ages. But to suggest that mankind lived in caves like animals and slowly learned to hunt, build, control fire, keep animals, and keep gardens until the race evolved into intelligent men is totally false and without **any** scientific support.

Being made in God's image gives man dignity and stature, and requires him to treat others with respect and kindness. It also requires him to live a godly life and to have a purpose to life. If man is only a refined animal, why not act like one? He can take what he wants when he wants it, satisfying all urges at anyone's expense. If man came from beasts, why not act like beasts?

However, man is not a little higher than animals, but a little lower than the angels according to the Psalms 8:5. Furthermore, the Psalmist said, "I am fearfully and wonderfully made" (Psalms 139:14).

Some sincere people honestly question creation because they honestly question the existence of a personal, all-powerful God. These people believe this because of various reasons: some were taught it in public schools, some at home, and others have been influenced by books, television and friends. They falsely assume that all intelligent people at least question the reality of God. And surely all scientists do! Not so, for many of the world's most famous scientists believe in a personal God. (And, until 1850 almost all scientists believed in God, and the most famous were Christians!)

Earlier I listed great scientists who were creationists; now I will provide names of scientists who believed in God. Of course, not all of those who believe in God have placed personal faith in Jesus Christ. Some of the better known recent scientists who believe in God are: Carl Jung; Albert Einstein; Sir John Eccles; A. E. Wilder-Smith, holder of three doctorates; Nobel Laureate in medicine and physiology, physicist Max Planck; Werner Heisenberg, world famous physicist; Robert Jastrow, founder/director of NASA's Goddard Institute of Space Studies; Henry Margenau, former editor of the **American Journal of Science;** Robert Sheldrake, former director of biochemistry at Cambridge University; and thousands of born-again Christians who hold

Ph.D. and Masters Degrees in areas of science, education, philosophy, theology, etc. Surely no sane person believes those people are incompetents or closed-minded fools.

All those scientists believe (or believed) in God, and believe that He could or did create this magnificent universe. The greatest scientist of all time, Isaac Newton, wrote, "This most beautiful system of the sun, planets, and comets, could only proceed from the counsel and dominion of an intelligent and powerful Being."[13]

The Bible says that God is responsible for everything we see, but evolutionists tell us He is not responsible because He does not exist. Then where did this incredible universe come from? Some scientists are willing to admit they honestly don't know. Scientist L. John concluded, "...the sad truth is that we do not know how the galaxies came into being."[14]

## THE BIG BANG

Unbelieving scientists pretend that God does not exist (they might as well pretend the sun doesn't shine), so they decide that creation could not have taken place and Genesis is not a scientific or historic fact. All right, then how did the universe get here? It **is** here! Trying to deal with **that** reality, they have posited the Big Bang Theory, but I believe the BBT takes more faith than creation! (The Bible **does** teach the Big Bang in Revelation where it reveals that the world will end with a Big Bang!)

We are supposed to believe that once upon a time (as all fairy tales begin) there was nothing, well yes there **was** something. There was space (and we creationists are supposed to **give** them that), and all the matter now in the universe was compressed into a sphere the size of a needle point![15] This is called, the "cosmic egg." Is it permissible to ask where **it** came from? Maybe it was laid by a cosmic chicken! And with time (where did **that** come from?) the egg exploded--producing the stars, planets and Earth! (Well, didn't I tell you the Big Bang was more miraculous than creation?)

According to a high school textbook, "...a fireball exploded 15 to 20 billion years ago. Then matter and energy spread outward in all directions, cooling as it expanded. After about 500,000 years, hydrogen gas formed. The gas collected into clouds which formed galaxies during the next half billion years. Now all that remains are galaxies and radiation. Within the galaxies, stars form and die and new ones form....Probably the most widely accepted theory for the origin of the solar system is the dust cloud theory. According to this idea, a dust cloud began to rotate....When the mass had swept up most of the material in an eddy, a planet was formed."[16] Proof? None!

While many assume the BBT is an accepted fact, the experts are not so convinced the theory is valid. J. Trefi says that one problem with the Big Bang is "how the galaxies could have formed in the time allotted for this process."[17] Leslie, author and scientist, agrees, saying, it "is hard to see how galaxies could have formed in a universe which is flying apart so fast."[18]

Astronomer Fred Hoyle said, "...an explosion merely throws matter apart, while the big bang has mysteriously produced the opposite effect, with matter clumping together in the form of galaxies."[19] Hoyle compares the BBT to an exploding bomb. The bomb explodes but doesn't do any damage until it hits an object, but with the BBT there was nothing in space to hit! After constant expansion, the matter should have dissipated, but we are to believe the opposite.

How did order come out of an explosion? Does that happen if a small bomb goes off inside a television set? Why and how could it happen in the universe? Leading British astronomer Paul Davies, wrote, "The greatest puzzle is where all the order in the universe came from originally."[20] One does not have to be a scientist to understand that simple principle.

If the universe is the result of an explosion, how does it run like a Swiss clock?--a comparison made by astronomer Johannes Kepler whose laws describe planetary orbits. Why, if the planets resulted from the big bang, do two planets revolve backward, (Venus and Uranus) and why do at least six moons (out of 60 in our solar system) rotate around their planets in a direction opposite to all the other moons?[21]

How could the same explosive thrust produce objects revolving in different directions? And how could an explosion produce a universe that runs as well as a clock?

Davies, in a **New Scientist** article wrote, "Everywhere we look in the universe, from the far flung galaxies to the deepest recesses of the atom, we encounter order...."[22] Nobel Prize winner, Max Planck agrees: "There is evidence of an intelligent order of the universe."[23] Einstein concurs, suggesting that the "high degree of order" was somewhat of a "miracle."[24]

Well, the Big Bang has started to fizzle! Astronomer Hoyle says that a "sickly pall now hangs over the big bang theory."[25] The Big Bang has fallen with a big bang! Eminent scientists who reject the BBT include Nobel Prize winner Hannes Alfven, astronomer Sir Fred Hoyle, astronomer Jayant Narlikar, astronomer N. Chandra Wickramasinghe, astronomer Geoffrey Burbidge, physicist Allen Allen, physicist Hermann Bondi, physicist Robert Oldershaw and physicist G. de Vaucouleurs.[26]

The Big Bangers have a Big Problem, as voiced by A. Krauskopt and A. Beiser: "A number of scientists are unhappy with the big bang theory....For one thing, it leaves unanswered the questions that always arise when a precise date is given for the creation of the universe: **Where did the matter come from in the first place?**"[27] (Emphasis added.)

This is also a concern of scientist Harlow Shapley who wrote: "In the very beginning, we say, were hydrogen atoms, of course, there must have been something antecedent, but we are not wise enough to know what. Whence came these atoms of hydrogen, these atoms, 20,000,000,000,000 (and 66 additional zeros) in number-- atoms that we now believe have been forged into the material make-up of the universe? What preceded their appearance, if anything? That is perhaps a question for metaphysics. The origin of origins is beyond astronomy. It is perhaps beyond philosophy, in the realm of the to us Unknowable."[28]

## MADE BY GOD

There are various theories as to the beginning of all things, but they can be distilled down to two theories: God created everything according to the Bible record or He did not. Under the column of "He did not" you can place the day-age theory, gap theory, theistic evolution, the Big Bang theory, etc. Either He did or He did not act according to the Scripture.

For those who believe that the Bible is the Word of God and that it means what it says, there is not an iota of doubt: God created the universe and everything therein in six 24-hour days!

God very clearly tells us in John 1:1-3, "In the beginning was the Word, and the Word was with God, and the Word was God. The same was in the beginning with God. All things were made by him; and without him was not anything made that was made." That passage is very clear, isn't it? **All** things means **all** things. Now, either believe God or not, but don't play around with the Truth.

In Acts 17:24 Luke tells us again that: "God that made the world and all things therein, seeing that he is Lord of heaven and earth, dwelleth not in temples made with hands." That is an affirmative statement that leaves no room for doubt: "God that made the world..." Did He or did He not? If not, the Bible is untrue, unreliable and unnecessary.

The Apostle Paul also tells us plainly that God is the Creator in Hebrews 1:10: "And, Thou, Lord, in the beginning hast laid the foundation of the earth; and the heavens are the works of thine hands." So the heavens are the works of God's hands; however, evolutionists want to make the heavens and the Earth the works of the god of chance and accident.

Now, since God created everything, you and I should react to His creative activity. Revelation 4:11 says: "Thou art worthy, O Lord, to receive glory and honour and power: for thou hast created all things, and for thy pleasure they are and were created." Here we discover that God not only created everything, but He created everything for His own pleasure. That verse is as true as John 3:16. If Christ died to save us then God created "all things."

When God finished with His creation work, He kept the control of everything for Himself. He keeps the universe running smoothly, planets in their orbits and the stars twinkling in the heavens. That truth is revealed in Colossians 1:16-17: "For by him were all things created, that are in heaven, and that are in earth, visible and invisible, whether they be thrones, or dominions, or principalities, or powers: all things were created by him, and for him: And he is before all things, and by him all things consist."

Do you get God's message? He keeps saying that **all** things were created by Him. Wonder why people make it so complicated? It is called our "sin nature." We have a natural tendency to rebel against the truth. Human nature prefers darkness to light, as we discover in John 3:19: "And this is the condemnation, that light is come into the world, and men loved darkness rather than light, because their deeds were evil."

**That** is the reason men refuse to believe that God created everything in six literal days. Unregenerate men prefer to sit in the darkness and boast of their great knowledge and wisdom, but "professing themselves to be wise, they became fools" (Romans 1:22). What a tragedy! Evolution is a bad joke; a very bad joke with tragic results.

Pathetic, pompous evolutionists sit in darkness speculating, questioning, theorizing and guessing about the origin of origins while all about them are hummingbirds, butterflies, roses, sunsets, rainbows, and man--all stamped: Made by God!

# CHAPTER THREE

# TRYING TO SATISFY UNBELIEVERS

While most thinking people believe God created everything, there are some professing Christians who are trying to construct a bridge between the fraud of gradualism and the fact of Genesis. However, the two teachings are diametrically opposite each other and are irreconcilable. Those Christians who are trying to satisfy unbelievers are erecting a bridge of mist, and all who try to cross are destined to fall into a noxious swamp of disbelief, disobedience and dismay. Those sincere, but foolish Christians do irreparable harm to other Christians and produce fodder for unbelieving scientists.

Evolutionist William Provine, of Cornell University, maintains that people who try to hold to religious beliefs while accepting evolution "have to check [their] brains at the church-house door."[1] He is right, and they also must "check their Bibles" at the door because they surely aren't using them if they believe **any** form of macroevolution.

Christians who hold any form of molecules-to-monkey-to-man evolution with one hand, and to the Bible with the other are ridiculed by evolutionists, who usually have more respect for creationists than they have for accommodators.

Some Christians, trying to accommodate unbelieving evolutionists, have propounded various theories that are unscriptural and are out of step with Christian leaders throughout the ages. They explain to us that they are only trying to use reason to solve the conflicts between science and the Scripture, but we must always be careful that we don't use reason as an excuse for **treason**. When one refuses the clear teaching of the Scripture, he has become a traitor to His cause. Of course, I am aware that good, even great men can disagree on the interpretation of Scripture, but when men refuse to take the clear teaching of the Bible because it conflicts with some Professional Speculator, it is treason.

## GAP THEORY

One theory, held by some Christians, is called, the "Gap Theory." Most of us who are over forty-five years old were taught this theory in Christian colleges. The theory was made popular by the Scofield Reference Bible in the early 1900s after a Scottish theologian, Thomas Chalmers, proposed it in 1814. Basically, the theory says that God created the heavens, Earth, animals and even a pre-Adamic race maybe billions of years before Adam. Then a "gap" before Genesis 1:2 is supposed to tell of divine judgment that devastated His original creation. This time between verse 1 and 2 lasted billions of years, and provides all the necessary time for strata to be laid, fossils to be encased in rock, coal to form, etc. Then God restored everything with His six days of creation.

In addition to the following biblical problems, the believer in the Gap Theory has a major practical problem. How did the world exist and "evolve" for billions of years without the sun that was not created until the fourth day?

However, this theory is pure speculation without any scriptural justification. The Gap Theory is unscriptural, unscientific and unnecessary. Those who teach this misinterpretation tell us Genesis 1:2 really says, "and the earth became without form, and void," but exegetical evidence will not support that position. Professor M. Henkel of the Winona Lake School of Theology contacted twenty leading Hebrew scholars, and asked if there was any evidence for a gap between verses 1 and 2, and they all replied, "No."[2]

We are expected to believe that a perfect world became "without form, and void," because of God's judgment. (Many believe it was a result of Satan being cast from Heaven.) The phrase, "without form, and void" means that the creation was not yet finished. And while the Bible teaches that Satan was thrown from Heaven, it does not tell us that there was a judgment upon God's creation as a result. That is reading something into the text that is not there.

"Without form and void" does not mean "chaotic." These are Hebrew words, **tohu wabohu,** that mean "empty and unformed." The earth was not yet finished for man's habitation. There were no flowers, rivers, fruit trees, grass, etc. But doesn't the fact that darkness was upon the face of the deep indicate that sin and judgment had contaminated God's creation? Not at all, for you must remember that God created light and darkness. The Psalmist wrote: "Thou makest darkness, and it is night" (Psalms 104:20). Just as the day has various purposes, so the darkness of night has its purposes. And everything God created was very good.

Whenever God created **anything,** it was good. But

if, according to the proponents of the Gap Theory, God's creation had been cursed, millions of His creatures had died and formed into fossils over millions of years, and Satan was now the god of this world, how could the Scripture affirm in Genesis 1:31 that it was "very good"?

Furthermore, Romans 5:12 teaches that death and destruction resulted from Adam's Fall. (After reading so many evolutionary texts, I believe Adam must have fallen on his head.) Paul wrote: "Wherefore, as by one man [Adam] sin entered into the world, and death by sin; and so death passed upon all men...." Then Paul said in I Corinthians 5:21, "....by man came death..." But if death, destruction and decay happened between Genesis 1:1 and 1:2 then Paul was wrong! If death is a result of man's sin, then death had to follow man rather than precede him. No, the Gap Theory, while not a heresy, is a fairy tale for Christians. Believe it if you want, but it is not taught in Scripture.

When Adam sinned in the Garden, he lost his innocence and acquired a sinful nature that was naturally selfish, arrogant, dishonest, etc. He no longer permitted God to control his life and sought his own way. He hid from God and covered himself with leaves. Man is still hiding from God. Some are hiding in great cathedrals, some in large universities, and some behind altruistic activities.

Man, a sinner, had to develop a world view that would eliminate God and put himself on the throne. Man insists on control of his own life without interference from a personal God. Man can escape his dependence on God, but he cannot escape the consequences of that decision.

Aldous Huxley admitted this in his book, **Ends and Means**: "I had motives for not wanting the world to have a meaning; consequently assumed that it had none, and was able without any difficulty to find satisfying reasons for this assumption....The philosopher who finds no meaning in the world is not concerned exclusively with a problem in pure metaphysics, he is also concerned to prove that there is no valid reason why his friends should not seize political power and govern in the way that they find advantageous to themselves....

"For myself as, no doubt, for most of my contemporaries, the philosophy of meaninglessness was essentially an instrument of liberation. The liberation we desired was simultaneously liberation from a certain system of morality. We objected to the morality because **it interfered with our sexual freedom**; we objected to the political and economic system because it was unjust. The supporters of these systems claimed that in some way they embodied the meaning (a Christian meaning, they insisted) of the world. There was one admirably simple method of confuting these people and at the same time **justify ourselves** in our political and erotic revolt: we could deny that the world had any meaning whatsoever."[3] (Emphasis added.)

That is the most honest and perceptive statement I have ever read that shows the basis of evolution and humanism. Huxley was honest, but a hedonist.

His hedonism came out on a television show when he was asked why evolution was so readily accepted. He replied that evolutionists accepted Darwinism even without proof because they didn't want God to interfere with their sexual mores. That does explain a great deal doesn't it? They expelled God from the universe so they don't have to acknowledge they will give an account to that God!

## THEISTIC EVOLUTION

Augustine, Bishop of Alexandria, Egypt, taught that God created the world, not in six literal days as recorded in Genesis 1 and 2, but only potentially. In other words, God started everything, and evolution slowly developed what God intended to take place--theistic evolution. This is God-started, God-directed gradualism, but gradualism is not supported by Scripture nor by science, as I will discuss later.

Some Christians who hold this theory feel very comfortable with it because they think they have satisfied both science and the Bible, when they have actually satisfied neither! They are neither fish nor fowl. Either God created the world as per Genesis, or He did not. Theistic evolution is a very sick, weak compromise with unbelieving scientists, and those ungodly scientists have more respect for creati nists than they have for theistic evolutionists.

But, we are asked, couldn't God have started everything billions of years ago and overseen the evolutionary process until man raised up on his two legs and walked out of a cave? Sure, he "could have," but just because something is possible doesn't mean it is probable. (And even if something were probable, it doesn't hold that it actually happened.)

When you accept theistic evolution as an explanation for our present condition, you must be willing to go all the way with it. The creation was not finished on the sixth day, so creation is still continuing. Man is climbing higher and higher to his anticipated perfect, noble, and elevated state, contrary to the clear Bible teaching that man is going down.

Besides, scientists will not "permit" any supernatural "interference" in man's origin because it leaves the door open for personal accountability. It defies the evolutionist's doctrine of naturalism that assumes the universe is a closed system of material causes and effects which prohibits any interference by God.[4]

How can a Christian twist the Bible like a pretzel--trying to make it say what it doesn't say or mean what it doesn't mean? Furthermore, a sovereign God could create everything in a few days as easily as in a few billion years. And if it took God a few billion years to enact His creation instead of six literal days, why didn't He say exactly what He meant to say? Of course, Christians believe that He did make Himself very clear in Genesis concerning origins.

## DAY AGE THEORY

Another theory argues that each "day" in Genesis 1 and 2 means a long age--in order to provide evolutionists with their millions of years to explain away God's creative acts. Evolutionists believe that with enough time, anything can happen.

George Wald wrote: "Time is in fact the hero of the plot. The time with which we have to deal is of the order of two billion years. What we regard as impossible on the basis of human experience is meaningless here. Given so much time, the 'impossible' becomes possible, the possible probable, and the probable virtually certain. One has only to wait: time itself performs the miracles."[5] (Wald has recently backed away from that position.)

There are places when "day" does mean an indefinite period of time such as the "day of the Lord," but whenever the Bible speaks of a certain number of days, the meaning is always literal 24-hour days. And Genesis speaks of the first day, second day, etc. without any reason to believe the meaning should be anything other than the normal meaning.

The Hebrew words for "morning " and "evening" are each used over a hundred times in the Old Testament, and they are **always** used in the literal sense. Why try to make any other sense out of it? Was God confused as to what He wanted to convey, or are the critics confused?

Was Christ in the grave three days or three ages? Was Jonah in the belly of the whale three ages? (Of course, those Christians who feel a need to explain away Genesis also would feel a need to give some human explanation for Jonah's three-day excursion in the deep.) Had Lazarus been dead three days or three ages? So when Moses wrote in Exodus 20:11 that "in six days the Lord made heaven and earth," why try to make it anything but literal days? There is surely no textual justification to believe otherwise.

No one would ever attempt to make the days of Genesis 1 anything other than 24-hour days except to reconcile the Bible account of creation with evolution. But we are asked how could the days be normal days since the sun was not created until the fourth day? That's no problem for the sun doesn't shine for up to six months north of the Arctic Circle and south of the Antarctic Circle, and no one would agree that at the North or South Pole a day is equivalent to six months elsewhere.

The Jewish Talmudic literature comments on almost every verse of Scripture, yet in no place is it suggested that the days of Genesis one are anything but normal 24-hour days.

Moses used the Hebrew word, **yom** for day in Genesis 1:15 where he said: "God called the light **yom**, and the evening and the morning were the first **yom**." Here the "day" is defined as the "light" period in the succession of periods of "light" and "darkness." Remember that the reason the sun and moon were created was to rule the day and night.

Furthermore, if God had intended for the days to be long periods of time, He could have had Moses use the Hebrew word, **olam** meaning "long periods of time" instead of **yom**.

Hebrew scholars agree with fundamental Christians concerning the literal days of Genesis. Professor James Barr, a renowned Hebrew scholar and Oriel Professor of the Interpretation of Holy Scripture at Oxford University, said in a personal letter on April 23, 1984: "So far as I know there is no Professor of Hebrew or Old Testament at any world-class university who does not believe that the writer(s) of Genesis 1 through 11 intended to convey to their readers the ideas that (a) creation took place in a series of six days which were the same as the days of 24 hours we now experience; (b) the figures contained in the Genesis genealogies provided by simple addition a chronology from the beginning of the world up to later stages in the Biblical story; (c) Noah's flood was understood to be world-wide and extinguished all human and animal life except for those in the ark."[6]

Genesis 2:13 tells us that God "rested" on the seventh day after concluding His six days of creation. Now, did God "rest" a day or an age? And if it was an age, then how does that make a legitimate symbol for the Hebrews taking the seventh day as their Sabbath as God had instructed them in Exodus 20:10-11?

Furthermore, Adam was created on the sixth day, and lived in the garden the remainder of that day; then he lived through the seventh day, and was driven out of the garden on the eighth, ninth or tenth day. Did he live through parts of three or more different geological ages? If so, he would have lived at least five hundred thousand years![7] Now they lived a long time in those days, but not **that** long.

But there are other reasons the days in Genesis 1 and 2 were literal 24-hour days. On the third day God created grass, herbs and trees, and every student knows that plants discharge life-giving oxygen and absorb poisonous carbon dioxide. The oxygen discharged by the plants is used by animals and people who then throw off carbon dioxide that is used by the plants! (Just one of thousands of examples of phenomena that happened "accidently" over a period of millions of years according to the sooth-sayers.)

However, if the days were really ages, the plants could not have lived without carbon dioxide since animals were not created until the fith day. Furthermore, grass and trees (created on the third day) could not grow without light, and the sun did not shine until the fourth day. Did the world spin millions of years without sunlight? If the day was an age, the grass and trees could not have lived, so the days must have been literal 24-hour days.

The day age theory is simply another myth that attempts to permit evolution and creation to kiss each other without accepting the clear teaching of the Word of God. Why don't people believe the miracles of creation if they believe the miracles of the Gospels? If Genesis 1 and 2 are myths, why not John 3?

No, the day age theory, like theistic evolution and the gap theory, are so much pious claptrap (while organic evolution is **pretentious** claptrap).

We expect unbelieving scientists to ridicule special creation, but often Christian journalists do a "bitter" job than unbelievers! (Of course, often a "Christian" is identified as anyone with a dusty Bible on his coffee table and membership in **any kind** of church.)

## ACCUSATIONS

Donald Chittick astutely points out this sad fact in his, **The Controversy: Creation vs. Evolution**, "Even more sad is the situation in which Christians side with pagans in attacking biblical presuppositions. Creation and those who believe it come in for special attack by those within the church who have adopted theistic evolution. For example, from just three articles appearing in Christian publications receiving wide circulation within the Christian community, I find the following list of accusations hurled at those who believe in creation:

Creationists:
Have a hidden agenda (are dishonest)
Have a poor strategy for Christian impact
Are aberrational
Are cultic
Lack training in historical sciences (biology; geology)
Have little contact with scientific community

Appeal to uninformed audiences
Do not make their case in established scientific channels
Are not as honest as evolutionists
Are unable to dent scientific orthodoxy
Have a garbled message
Have a message that is just plain wrong
Have a message which often lacks accuracy
Have a message which often lacks balance
Magnify a particular view out of proportion
Are hyper-literalists
Sweep aside evidence
Present a caricature of the true Christian view
Are isolationist
Are inflexible
Are doing more harm than good
Abandon science
Are kibitzers of science
Help perpetuate unbelief
Commit logical fallacies
Use verbal chicanery
Misuse scientific theory
Misinterpret the Bible
Are guilty of the 'growing edge syndrome' logic
Shadowbox with straw men
Strain factual data
Abuse reason
Use circular reasoning
Narrowly read Scripture
Flagrantly ignore scientific evidence."[7]

Well,  now  you know what **Christianity Today**  and

**Eternity** think about you if you believe the Bible. Of course, you can find some of us who are guilty of some of the above, but it is outrageous for those magazines to suggest the above are symptomatic of Bible believing Christians. Those magazines are not known for their adherence to the veracity of Scripture, nor are they known for their balance. Sometimes they manage to rise to some heights of intellectual thought, but never to the height requiring additional oxygen!

In the following chapter I will seek to prove that we will be much safer, happier and scientifically correct if we simply accept the Bible for what it says after proper exegesis. God created all things by the word of His power in six literal days, and all the thumbsucking evolutionists (theistic or otherwise) in the world can't change that fact.

Mammoth--Extinct Elephant

# CHAPTER FOUR

# IS THE BIBLE RELIABLE IN EVERY DETAIL?

Often when men are unsure of their own arguments, they attack their opponents to draw attention away from their own weak, untenable position. Such is the case with Bible critics, especially evolutionists who have a vested interest in denying, denigrating and disproving the validity of the Word of God. In this chapter I will seek to prove that God is more reliable than His critics.

Men have known about germs since Louis Pasteur's work over one hundred years ago, but God warned about contagion more than 3,000 years ago! He warned people not to touch dead bodies in Numbers 19:11, 14, and 16, and He commanded lepers to be separated from the uninfected. He also told the Hebrews to wash their hands in running water. If the Bible were simply the work of men, how did Moses know to follow proper health procedures that have only been known for a little over a hundred years?

Some Christians believe the Bible is reliable in spiritual and moral matters, but not in historic and scientific matters, but that is a heretical view. If the Bible is not reliable in **all** areas, it is not reliable in **any** area. In recent years, even non-fundamentalist professionals have come to believe the Bible is trustworthy! This, however, is a well kept secret.

Archaeologist Nelson Glueck, professor at Hebrew Union College, said, "I have excavated for 30 years, with a Bible in one hand, and a trowel in the other. In matters of historical perspective, I have never yet found the Bible to be in error."[1] That is some admission from a professor at a Jewish college!

Professor George E. Wright, Professor of Semitics and Old Testament Studies at Harvard University, admitted that he had been wrong in telling his students that they could forget Moses in the Pentateuch![2]

Archaeologist William Albright, not a fundamentalist, has acquired a gradual respect for the truthfulness of the Bible. Albright states: "There can be no doubt that archaeology has confirmed the substantial historicity of Old Testament tradition."[3] The highly respected scholar also wrote: "Discovery after discovery has established the accuracy of innumerable details, and has brought increased recognition to the value of the Bible as a source of history."[4]

Albright added in another volume: "...there is scarcely a single biblical historian who has not been impressed by the rapid accumulation of data supporting the substantial historicity of patriarchal tradition."[5] Albright's "conversion" to Bible accuracy is a major coup for Bible Christians.

Millar Burrows of Yale University wrote: "Archaeology has in many cases refuted the views of modern critics. It has shown in a number of instances that these views rest on false assumptions and unreal, artificial schemes of historical development."[6]

Archaeologist Merrill Unger agreed with the above quotation when he said: "Old Testament archaeology has rediscovered whole nations, resurrected important peoples, and in a most astonishing manner filled in historical gaps, adding immeasurably to the knowledge of biblical backgrounds."[7] Creation Scientists have nothing to fear from the next spade of dirt of the archaeologists. Nothing!

## BIBLE CRITICS WRONG

We were told in university classes that camels did not exist in Palestine during the time of Abraham; however, we now know they were wrong. Camel teeth and bones have been unearthed there and figurines dating back to 3000 B.C. show camels in the land before Abraham!

We were also taught that lions were not to be found in the Middle East so the tale of Daniel in the lions' den was just that, a tale. But a black basalt statue of a lion was found at Babylon. Its age was from the time of Daniel! While lions were not natural to the area, they were imported for kings to hunt.

Unbelievers have ridiculed the giants recorded in the Bible, but they only show themselves to be uninformed fools. Many giants have lived among us even in recent times. **The London Times**, of June 14, 1880 reported a reception at the Royal Aquarium hosted by Chang of Pekin who was eight feet two inches tall. Would he qualify as being a giant?

What about Captain Bates of Kentucky who was escorted around London in 1871? He was eight feet tall, **and so was his wife, Anna Swan**! They made a big splash in London society.

Charlemagne, the ruler who consolidated Europe in the eighth century was almost eight feet tall, and could squeeze three horseshoes together with his hands.

How about Pliny, who reported that Pusic and Secundilla, living during the reign of Augustus, were over ten feet tall! Is the record of Pliny to be believed in other areas, but not the biblical report, because it doesn't fit into the preconceived opinions of the Bible skeptics?

Then there was Patrick Carter who was eight feet, seven and one-half inches tall who died in Clifton, Bristol, in 1802. This Irish giant's hand is preserved in the Museum of the College of Surgeons.

John Frederick, Duke of Brunswick, was eight feet six inches tall; John Middleton, born in the reign of James I, was nine feet three inches tall; Louishkin, the Russian giant and drum major of the Imperial Guards, was eight feet five inches in height; the porter of Charles I, William Evans, who died in 1632, was eight feet tall; and finally, Gabara, the Arabian giant, was nine feet nine inches tall according to Pliny.[8]

Christians don't need these historical examples to be convinced that the Bible is true, but history does show that Bible critics are usually very uninformed people.

The story of Joseph and the seven years of famine was pronounced a myth because the idea of the Nile River not overflowing was unthinkable, but in recent years, tablets and monuments were discovered that recorded the story in detail!

For many years the story of the Queen of Sheba's visit to Solomon was considered an unreliable folk tale handed down from generation to generation; however, excavations at Marib have proved that Sheba was one of four important Arabian kingdoms. Our critics were wrong again!

Critics have questioned the enormous wealth of

Solomon as recorded in the Bible, but the spade of the archaeologist has uncovered his garrison cities of Megiddo, Hazor and Gezer. Solomon's stables, sea port and copper refineries have been discovered, proving he sat on an extended and wealthy throne.

Jericho has long been an issue of dispute for the archaeologists, and has been covered extensively in professional archaeological magazines. Some even question Joshua's leadership and his success in conquering the pagan tribes that inhabited the land at the time. The big dispute is whether Joshua took the city and if the walls fell **out**, as recorded in the Bible.

Dr. John Garstang, director of Archaeology in Jerusalem, discovered that the walls did fall outward even though they were 15 feet high and 10 feet thick. He also discovered, from pottery and scarab evidence, that the city had been destroyed about 1400 B.C., about the time of Joshua![9]

I have been to the Jericho dig more than ten times and can attest to the enormous walls, towers and fortifications. The modern city of Jericho is within walking distance of the dig.

Bible critics have ridiculed the biblical account of the Hittites who are mentioned 46 times in the Bible, but no where else in history. So, of course, they assumed that the Hittites were fictitious. Wrong again, for archaeologists have discovered that the Hittites were one of the leading nations of ancient times! There are now hundreds of references that prove the existence of the Hittites!

The Exodus of the Jews and the conquest of Canaan by the Israelites in about 1400 B. C. were thought to be patriotic folk tales passed on to Jewish children, but 377 clay tablets were found in Egypt that described those events. Those tablets belonged to Pharaohs Amenhotep III and IV who ruled in the time of 1500-1400 B.C.[10]

The Assyrian king, Sargon, as mentioned in Isaiah 20:1, was fictitious, according to liberal Bible critics. But his palace at Khorsabad has been uncovered, and the very battle referred to by Isaiah was mentioned on an inscription! Critics were wrong again and the Bible right--again.

Nebuchadnezzar was supposed to be an unreal character dreamed up in the fertile mind of Daniel; however, when the southern palace of Babylon was excavated, proof was found that Nebuchadnezzar had lived and had a part in the construction of the palace as recorded in Daniel 4:30!

Every Bible critic has trouble swallowing the story of Jonah (because they don't believe in the veracity of the Bible). It was thought that no fish could swallow a man, but now we know that is not correct. We now know that the sperm whale, the white shark and the Rhinodon Shark can and have swallowed men!

For more than twenty years I have been using the example that was published in the **Literary Digest** in 1891 about James Bartley who was a sailor on the **Star of the East**. The ship was near the Falkland Islands when the lookout spied a sperm whale about three miles away. Two boats were lowered and the whale was harpooned. The injured whale lashed one boat with its tail, throwing two men into the water. One man was drowned, and James Bartley was presumed drowned; however, the whale swallowed him while he was thrashing in the water. The next day, the whale was cut open, and James was discovered alive  although unconscious. After treatment in a London hospital, he recovered fully from the ordeal, except that his skin was tanned by the gastric juices. There are many similar examples of men living through such an experience. So, why not Jonah?

Bible critics love to point out "problems" in the Bible, thereby seeking to dig out the very foundations of our faith and justify their unbelief. Of course, while there are some passages difficult to explain, the Bible is the very Word of God, without error or contradiction. The problems are not in the Bible but in the sinful hearts of men.

Many "contradictions" are simply the critics' poor reading of the Scripture, their abysmal ignorance of history or their misunderstanding of manners and customs of the Middle East during the Bible era.

A good example is the two "different" versions of creation in Genesis one and two. One does not have to be too bright to realize that one account is specific and the other is general. In Genesis 1:1--2:4a, God gives us a wide angle view of His creation, and in Genesis 2:4b-24, He zooms in on the sixth day providing us the details that were left out of the general view. This is a very common practice for writers and speakers, and it is incredible that intelligent men would use something so elementary as a criticism when no criticism is justified. It is called "desperation."

Another similar "problem" is the inscription on the cross of Christ. During television talkshows, opponents have thrown this "contradiction" at me as if they were really causing me great anguish. I usually point out that this is a sophomoric criticism, and that they could find far more difficult passages if they were really informed.

The Gospels do seem to give a different wording of the sign over the cross, but it is very simple to explain. No one gospel quotes the entire inscription that appeared on the cross. All four gospels have, "The King of the Jews." Luke adds the further words, "This is." Matthew quotes the name, "Jesus" and John adds "of Nazareth." Combining the entire inscription, it reads, "This is Jesus of Nazareth the King of the Jews." A very simple problem to solve, but the "problem" is never in the Bible, but in the hearts of men.

Bible critics have accused Bible authors and Christians in times past, of teaching a flat Earth that rested on giant pillars, but of course, that is a false accusation used to produce lack of confidence in the Bible. Some Roman Catholic leaders promoted such a teaching, furthered by artists, but Christian leaders did not preach that error.

We read in Isaiah 40:22, "It is he that sitteth upon the **circle** of the earth, and the inhabitants thereof are as grasshoppers, that stretcheth out the heavens as a curtain, and spreadeth them out as a tent to dwell in." Did you notice the "circle of the earth"? Well, there is no flat earth there.

The Bible describes a spherical Earth suspended in space, as Proverbs 8:27 says, "When he prepared the heavens, I was there; when he set a **compass** upon the face of the depth." Obviously, there is no flat earth there.

The words, "compass" and "circle" are both translations of the same Hebrew word, **chuwg**. Either circle or sphere is a good translation of the word. Bible critics are wrong. The Bible taught that the Earth was a sphere thousands of years before any scientist did! The big surprise is that anyone could believe in a flat earth when the Bible is so clear that it was otherwise. Most of the other alleged

Bible discrepancies can be answered as simply.

God further "clears the air" on this matter in Job 26:7 where He said, "He stretcheth out the north over the empty place, and hangeth the earth upon nothing." The original word means, "nothing whatever." Note that the verse speaks of a northern "empty space." Of course, Job would not have known that without Holy Spirit inspiration.

Could this "empty space" be the "hole in space" that was recently discovered by astronomers? M. Mitchell Waldrop wrote, "The recently announced 'hole in space,' a 300 million-light-year gap in the distribution of galaxies, has taken cosmologists by surprise..."[11] Of course, God was not surprised since He distributed the galaxies!

The Bible is true and reliable, and has stood the test of time. Christianity has been pronounced dead many times by Ingersoll, Voltaire, and yuppie scientists, among others. They have preached its funeral, held its wake, engraved its tombstone, but there is no corpse! It is alive and well!

## SCIENCE SUPPORTS BIBLE

Now I will deal with the biblical proof for belief in special creation. Earlier in this chapter I sought to substantiate the validity of the Bible, and how the ranting and ravings against it are not valid, honest or reasonable.

The Bible is the unique, verbally inspired Word of God, not simply one of many "sacred" texts. The other "sacred" texts are riddled with errors and outrageous teachings such as the writings of the Hindus where the moon is said to be 50,000 leagues higher than the Sun and shines by its own light, and the Earth is said to be flat and composed of seven layers. One layer is of honey, another of sugar, another of butter and still another of wine! This mass (or is it mess?) is borne upon the heads of numerous elephants who, when they stumble, produce earthquakes. The feet of the elephants are resting on the shell of an immense tortoise, and the tortoise on the coil of a great snake. No wonder they stumble! How would you like to be trusting that book to provide the message of redemption?

Bible information about astronomy is a proof of creation. Astronomers have told us that there are an estimated 10 million billion billion stars, while Ptolemy suggested the number was 1,058 and Kepler counted 1,005. However, we know that it is impossible to number the stars. God said in Jeremiah 33:22, "As the host of the heaven cannot be numbered."

After Lord Rosse invented his telescope, it was discovered that there was a huge empty space in the north. That was a great discovery, but Job wrote thousands of years earlier in Job 26:7: "He stretcheth out the north over the empty space."

The Bible also teaches in Job 26:7 that "He hangeth the earth upon nothing." Until Copernicus, most of mankind believed the Earth rested on a firm foundation. Job knew that the Earth was not on a firm foundation long before scientists. How did he know that except by Divine inspiration?

Furthermore, the Bible teaches that the Earth, is round as recorded in Isaiah 40:22: "It is God who sitteth upon the circle of the earth." How was it possible for men to write of scientific principles thousands of years before science had discovered them? It is called inspiration of Scriptures.

Ancient people believed the moon was larger than the Earth or the Sun and was its own light source; however, Bible readers knew otherwise. Genesis 1:16 reveals, "And God made two great lights; the greater light to rule the day, and the lesser light to rule the night..."

Maybe these astronomical Bible truths are one reason Dr. C. A. Chant, Professor of Astrophysics at Toronto University declared, "I have no hesitation in saying that at least ninety percent of astronomers have reached the conclusion that the universe is not the result of blind law, but is regulated by a great intelligence."[12]

Renowned astronomer Robert Jastrow said as much in his 1982 book where he wrote, "Astronomers now find they have painted themselves into a corner because they have proven, by their own methods, that the world began abruptly in an act of creation to which you can trace the seeds of every star, every planet, every living thing in this cosmos and on the earth. And they have found that all this happened as a product of forces they cannot hope to discover."[13] I suggest you read that paragraph again!

Zoology supports the Bible as revealed in Job 18:7, where it says, "Which no fowl knoweth and which the vulture's eye hath not seen." This teaches that the vulture locates his prey by sight and not by scent. Critics scoffed at this "mistake," but it was no mistake. We now know that the vulture's eye is a powerful telescope with which he discovers his prey.

The Bible and meteorology are also in harmony, as seen in Ecclesiastes 1:6-7: "The wind goeth toward the south, and turneth about unto the north; it whirleth about continually, and the wind returneth again according to his circuits. All the rivers run into the sea; yet the sea is not full; unto the place from whence the rivers come, thither they return again."

Job revealed an incredible truth long before scientists discovered it. He wrote in Job 26:8, "He bindeth up the waters in his thick clouds; and the cloud is not rent under them."

It was not until Galileo's time (1630 A.D.) that men knew the winds had regular circuits and rain clouds were only water taken from the rivers and seas by evaporation.

No scientist, before Galileo, knew that the air had weight, but Job knew it thousands of years ago, when he said in Job 28:25: "To assign to the wind or air...its weight." It might be helpful if scientists spent more time reading the Bible and less time in scientific journals--journals that keep changing with new "discoveries."

Of course, the Bible is true whether or not well known scholars accept it; God's existence does not depend upon our belief! Those who accept the Bible as true do not need science to confirm it; however, it is enjoyable to see the unbelieving scientists squirm when we don't simply quote the Bible but throw scientific facts at them. That's what I will do in the remainder of this book. I enjoy beating skeptics over the head with Truth, proving that the Truth of the Bible is reliable and the religion of evolution is a false science and a false religious system.

# CHAPTER FIVE

# IS EVOLUTION A RELIGION?

In almost every creation-evolution debate the scientist states that scientific creationism, because it is religion and not science, must never be taught in the public school--schools that have been taken over by the teachers' unions. However, it is also true of evolution since it is simply a humanist, materialistic religion that seeks to explain man's origins. (It's not a very good religion; doesn't have any holidays except April 1, although most evolutionists get rather pious on the anniversary of the founding of the ACLU.)

Renowned astronomer Robert Jastrow concluded that one must start with faith whether as an evolutionist or a creationist. He wrote, "Perhaps the appearance of life on the Earth is a miracle. Scientists are reluctant to accept that view, but their choices are limited. **Either** life was created on the Earth by the will of a being outside the grasp of scientific understanding, **or** it evolved on our planet spontaneously, through chemical reactions occurring in nonliving matter lying on the surface of the planet.

"The first theory places the question of the origin of life beyond the reach of scientific inquiry. It is a statement of faith in the power of a Supreme Being not subject to the laws of science.

"The second theory is also an act of faith. The act of faith consists in assuming that the scientific view of the origin of life is correct, without having concrete evidence to support that belief."[1] (Emphasis his.)

So one must start with an assumption, and that requires faith; however, that does not weaken one's position. It's what he does with the assumption that weakens or strengthens his position. No human knows everything, so he must make assumptions. Evolutionists are dishonest when they deny their "faith." They do not begin with facts but with faith (and usually end with a farce and often a fraud). Neither evolution nor creation has been observed, and neither can be observed in a laboratory. There must be a working assumption or working hypothesis.

## A MYTHICAL RELIGION

Evolution is not science but science fiction. It is scientism, a mythical religion. The inscription on the National Academy of Sciences building reveals that fact: "To science, pilot of industry, conqueror of disease, multiplier of the harvest, explorer of the universe, revealer of nature's laws, eternal guide to truth."[2] Many fundamentalists don't talk about **God** with such reverence!

Renowned historian of science and philosopher, Marjorie Grene, wrote, "It is as a religion of science that Darwinism chiefly held, and holds men's minds....The modified but still characteristically Darwinian theory has itself become an orthodoxy, **preached by its adherents with religious fervor**, and doubted, they feel, only by a few muddlers imperfect in scientific faith."[3] (Emphasis added.)

Note that Grene admits that evolution "holds men's minds" and has become "an orthodoxy" and that orthodoxy is being preached "by its adherents with religious fervor." Would that more Christians had the dedication and enthusiasm that many evolutionists have.

Some evolutionists have almost made Darwin god as admitted by evolutionist Colin Patterson, senior paleontologist at the British Museum of Natural History: "Just as pre-Darwinism biology was carried out by people whose faith was in the Creator and His plan, post-Darwinian biology is being carried out by people whose faith is in, almost, the deity of Darwin."[4]

I surely would not want to base my eternal destiny on the speculations, guesses and wild theories of Charles Darwin.

Well-known scientist Heribert Nilsson of Lund University emphasized the religion of evolution when he wrote: "My attempts to demonstrate evolution by an experiment carried on for more than 40 years have completely failed...It is not even possible to make a caricature of an evolution out of paleo-biological facts....The idea of an evolution rests on **pure belief**!"[5] (Emphasis added.) That is religion!

If you want to impress people, and especially if your argument is anemic, you could start with, "Science has proved..." or "Scientists believe..." These statements are used to help support an unsupportable argument. They are used to preempt a rebuttal, and you can get away with almost any statement when you are dealing with the unsophisticated, unenlightened and uneducated.

We must learn to ask, "But is it **true**? How do you

**know** it's true? What is your **evidence**? Are you implying that **all** scientists agree with that position? Is it a **fact** or is it your interpretation of a fact?" There is a difference you know.

B. Leith admitted, "It is therefore of immediate concern to both biologist and layman that Darwinism is under attack. The theory of life that undermined nineteenth-century religion has virtually become a religion itself and in its turn is being threatened by fresh ideas.[6]

Evolutionist Julian Huxley admitted that his belief was the religion of "evolutionary humanism."[7] Well, at least Huxley was honest, unlike most evolutionists.

Harold Urey, Nobel Prize laureate said the same thing: "[A]ll of us who study the origin of life find that the more we look into it, the more we feel it is too complex to have evolved anywhere. We all believe as **an article of faith** that life evolved from dead matter on this planet. It is just that its complexity is so great, it is hard for us to imagine that it did."[8] (Emphasis added.)

Biochemist G. A. Kerkut agreed: "It is therefore a matter of faith on the part of the biologist that biogenesis did occur and he can choose whatever method of biogenesis happens to suit him personally; the evidence for what did happen is not available."[9] So, he admits that what he believes about the beginning of life is a matter of faith. Sounds like religion to me.

The renegade Jesuit scientist-priest, de Chardin's statement is a religious pronouncement: "Is evolution a theory, a system, or a hypothesis? It is much more--it is a general postulate to which all theories, all hypotheses, all systems must henceforth bow and which they must satisfy in order to be thinkable and true. Evolution is a light which illuminates all facts, a trajectory which all lines of thought must follow--this is what evolution is."[10]

That is one of the most religious statements I've ever read. **All** theories, **all** hypotheses and **all** systems must bow before evolution, and evolution is **light**! But it is not light, but darkness. Only God is light. His statement is a revelation of pure paganism.

## A NEW RELIGION

Julian Huxley, avid evolutionist, announced the Religion of Evolution in Chicago when he said, "In the evolutionary pattern of thought there is no longer either need or room for the supernatural. The earth was not created, it evolved. So did all the animals and plants that inhabit it, including our human selves, mind and soul as well as brain and body. So did religion....

"Finally, the evolutionary vision is enabling us to discern, however incompletely, the lineaments of the **new religion** that we can be sure will arise to serve the needs of the coming era."[11] (Emphasis added.) He said the "R" word, didn't he?

Huxley co-authored a book with British evolutionist, Jacob Bronowski where they wrote, "A religion is essentially an attitude to the world as a whole. Thus evolution, for example, may prove as powerful a principle to coordinate man's beliefs and hopes as God was in the past."[12]

Evolutionists teach that religion evolved, along with everything else, as a result of man's fear of nature, the future, death, etc. so **man created God** rather than God creating man! And Huxley and Bronowski suggest that the religion of evolution may be successful in codifying man's "belief and hopes" as religion did in man's early history.

Why do intelligent people still accept evolution when it has no facts to support it (as will be discussed in following chapters)? It is because it has become their religion! The foundation is not facts; it is faith!

British paleontologist Colin Patterson, an evolutionist for twenty years, admitted that he, like most scientists, had assumed that an evolutionary view of nature had replaced an irrational one [creation], but in 1980 that changed for Patterson. He admitted, "Then I woke up and realized that all my life I had been duped into taking evolutionism as revealed truth in some way." He said that he had experienced "a shift from evolution as knowledge to evolution as faith."[13]

The following statement is one of the most religious ones you will ever read, and it was made by Dr. George Wald, winner of the Nobel Prize in Science: "When it comes to the origin of life on this earth, there are only two possibilities: creation or spontaneous generation (evolution). There is no third way. Spontaneous generation was disproved 100 years ago, but that leads us only to one other conclusion: that of supernatural creation. We cannot accept that on philosophical grounds (personal reasons); therefore, we choose to believe the impossible: that life arose spontaneously by chance."[14] **That**, my friend, is religion.

Wald, admits that evolution is impossible, but he believes it anyway! At least those of us who believe in miracles, believe that an omnipotent God performs those miracles by suspending natural laws. Evidently, Wald and other scientists would rather appear to be religious fools than religious fanatics. I'd rather be a "fanatic."

Evolutionist G. G. Simpson put his religion in proper perspective when he said, "It is inherent in any acceptable definition of science that statements that cannot be checked by observation are not really about anything...or at the very least, they are not science."[15] If the teachings are not science, as admitted by various evolutionists, then what are they? This is religion, pure and simple.

Famous British biologist and evolutionist L. H. Matthews wrote the introduction to the 1971 edition of Darwin's **Origin of Species**, and wrote of his faith when he said: "Most biologists accept it [evolution] as though it were a proven fact, although this conviction rests upon circumstantial evidence; it forms a satisfactory faith on which to base our interpretation of nature."[16] Well, at least he was honest in calling evolution a "satisfactory faith," and that statement is found in the introduction to Darwin's book!

Matthews also admitted: "The fact of evolution is the backbone of biology, and biology is thus in the peculiar position of being a science founded on an unproved theory-- is it then a science or faith? Belief in the theory of evolution is thus exactly parallel to belief in special creation--both are concepts which believers know to be true but neither, up to the present, has been capable of proof."[17]

Other than his "fact of evolution being the backbone of biology," the statement is generally true; however, Christians accept the Bible as complete proof and scientific evidence as supporting proof. We can do without the science, but not without the Bible!

Physicist Dr. Wolfgang Smith agrees that evolution is religion. He wrote, "The point, however, is that the doctrine of evolution has swept the world, not on the strength of its scientific merits, but precisely in its capacity as a Gnostic myth. It affirms in effect, that living beings created themselves, which is in essence, a metaphysical claim....Thus, in the final analysis, evolution is in truth a metaphysical doctrine decked out in scientific garb."[18]

Dr. Smith characterizes evolutionists as priests rather than objective seekers of scientific truth. Evolution is a religion, a cult that doesn't require church attendance, godly living or tithing.

More and more rabid evolutionists are admitting that evolution is a religion based on faith not facts. Evolutionist Kenneth Hsu quoted evolutionist Paul Feyerabend approvingly when he wrote, "...I agree with him that Darwinism contains 'wicked lies'; it is not a 'natural law' formulated on the basis of factual evidence, but a *dogma*, reflecting the dominating social philosophy of the last century."[19] (Emphasis added.)

L. T. More of the University of Cincinnati, wrote, "Our faith in the doctrine of evolution depends upon our reluctance to accept the antagonistic doctrine of special creation."[20]

More also confessed: "The more one studies paleontology, the more certain one becomes that evolution is based on faith alone; exactly the same sort of faith which is necessary to have when one encounters the great mysteries of religion."[21]

Webster's Dictionary defines "faith" as an "unquestioning belief, specifically in God, religion, etc." The "etc." covers the religion of evolution. The definition of "religion" is "a cause, principle, or system of beliefs held to with ardor and faith." **That**  also describes evolution.

## THE TRUTH

Mankind needs something authoritative in which he can believe, something that never changes and something that can be defended against outrageous attacks. He has that in the immutable Word of God. And that Book is not only reliable for spiritual matters but in **all** other matters as well. That includes the origin of life.

Notice that God makes it very clear that He created **everything** as is recorded  in Genesis 1  and 2 and confirmed in Nehemiah 9:6: "Thou, even thou, art Lord alone; thou hast made heaven, the heaven of heavens, with all their host, the earth, and all things that are therein, the seas, and all that is therein, and thou preservest them all; and the host of heaven worshippeth thee."

God seems to be trying to make doubly sure that we understand that He made the Earth and everything on it, the seas and everything in them, and the heavens and everything in them! God seems to be speaking in advance to those who will refuse to take Him at His Word.

In Isaiah 45:18 God said, "For thus saith the Lord that created the heavens; God himself that formed the earth and made it; he hath established it, he created it not in vain, he formed it to be inhabited: I am the Lord; and there is none else."

So the Bible does teach special Creation. Pompous men may try to twist and pervert it to teach otherwise, but everyone will admit that the Bible does say that "In the beginning God created the heaven and the earth." Why not take God at His Word? After all, isn't He worthy of our trust?

Evolutionists imply that all sane, intelligent people accept their fairy tale concerning origins, but there are hundreds of creationists with advanced degrees (over 1000 in one professional group alone) who are vocal witnesses to the dishonesty in their accusations. Evolutionists can't deal with the issues involved, so they slander their opponents. Instead of attacking our positions, they attack our persons. Very sophomoric!

Some of the greatest scientists of all time were creationists. Men like Joseph Lister, Louis Pasteur, Isaac Newton, Johann Kepler, Robert Boyle, Georges Cuvier, Michael Faraday, Ambrose Fleming, Lord Kelvin, Louis Agassiz, James Simpson, William Ramsay, James Maxwell, Gregor Mendel, Blaise Pascal, and many others were creationists. These men are among the greatest scientists who ever lived.

Dr. Dean H. Kenyon, Professor of Biology at San Francisco State University, is a creationist who says that creationism is scientific and reasonable: "It is my conviction that if any professional biologist will take adequate time to examine carefully the assumptions upon which the macroevolutionary doctrine rests, and the observational and laboratory evidence that bears on the problem of origins, he/she will conclude that there are substantial reasons for doubting the truth of this doctrine. Moreover, I believe that a scientifically sound creationist view of origins is not only possible, but is to be preferred over the evolutionary view."[22]

Obviously all informed, educated people do not accept evolution as a fact, and all creationists are not backwoods snake handlers. When we face off in debate with evolutionists, they usually end up looking like bungling incompetents. In fact, occasionally scientists and philosophers who are non-believers and even critics have come to the creationist position!

Agnostic physicist Lipson made an amazing confession in his book: "I think, however, that we must go further than this and admit that the only acceptable explanation is **creation**. I know that this is anathema to physicists, as indeed it is to me, but we must not reject a theory that we do not like if the experimental evidence supports it."[23] (Emphasis added.) What an admission from the enemy!

And no one ever accused John Stuart Mill of being a Bible-spouting Christian, but listen to him give his testimony: "Leaving this remarkable speculation [natural selection] to whatever fate the progress of discovery may have in store for it, I think it must be allowed that, in the present state of our knowledge, the adaptations in Nature afford a large balance of probability in favour of creation by intelligence. It is equally certain that this is no more than a probability...."[24]

Astronomer Hoyle seems to continue Mill's thought: "Rather than accept that fantastically small probability of life having arisen through the blind forces of nature, it seemed better to suppose that the origin of life was a deliberate intellectual act. By 'better' I mean less likely to be wrong."[25]

Attorney W. R. Bird suggests very astutely that it is a mistake to state categorically that the incredibly improbable cannot happen, or that the more probable must happen![26]

There is a battle going on between creationists and evolutionists, and some honest evolutionists are being "converted" to our position without being converted to Christ. That is a tragedy.

Scientists have their prejudices, ego problems, and religious bias like everyone else, and they go into the creation-evolution debate with a bias against God and the Bible. Many of them never give a serious consideration to the facts as presented in this book. They are standing on sinking sand and building on a foundation that is cracking on all sides. But, to their dying day, many of them must play the intellectual and deny the Book that begins, "In the beginning, God created the heaven and the earth."

Zealous evolutionists are standing guard over their religion like eunuchs guarding a harem of graying, plump and age-weary concubines. They have a religious devotion to evolution, the Big Lie, and they worship it as surely as pagans bow down to dumb, lifeless, powerless idols.

All evolutionists should have the following poem on their office wall:

> As I was sitting in my chair,
> I knew it had no bottom there,
> Nor legs, or back, but I just sat,
> Ignoring little things like that.[27]

**Darwinian** evolutionists are sitting on nothing but faith while they often disdain religion! They are followers of an apostate preacher whose speculations have been elevated to science but they are not truth-sayers but sooth-sayers! In reality, they are the high priests of their nihilistic religion that says, "In the beginning was hydrogen." But I serve the God who created the hydrogen!

# CHAPTER SIX

## DID DARWIN FIRE A BLANK?

In 1859 Charles Darwin fired a shot heard around the world, but Charlie fired a blank with the publication of his **Origin of Species by Means of Natural Selection or the Preservation of Favoured Races in the Struggle for Life**, (in which he did not deal with the origin of species)! Evolutionists love to boast how his book sold out the first day it was released to the public; however, they **never** tell you that they only printed 1250 copies! Big deal!

Darwin shocked the world with "his" theory that all living creatures came from a single ancestor without any help from a sovereign God. It all happened, so the story goes, verrrry slowly over millions of years, but some of us think that Federal law should require every lecture on evolution to begin, "Once upon a time." After all, all fairy tales begin that way, and evolution is a fairy tale for adults, but usually foisted off on children.

The world took Darwin's **guess** to be gospel, and accepted him as a great authority, although his education at Cambridge was for the ministry! His father and grandfather were physicians, and Charles sought to follow in their footsteps, but failed to make it because he spent his time with a worthless crowd. Charles could not get excited about his studies, but he was always excited about rat catching. His father scolded him with the words, "You care for nothing but shooting, dogs and rat catching, and you will be a disgrace to yourself and all your family."[1]

Well, his father's prediction turned out to be true! It would have been better for Darwin and for the world if he had spent his life shooting and catching rats!

Charles wasn't too concerned about his future since he knew he would inherit his father's fortune, but he had to have some profession. He, with the help of his father, entered the ministry! In those days, the dull and unmotivated young men went into the ministry; today they go into public education, a fact known on every college campus in America.

Darwin entered Cambridge University in 1828, and finally graduated with a theology degree! Yes, the man that is often regarded as a great scientist by uninformed and adulating people was an apostate minister! After Cambridge, Darwin planned to enter the ministry somewhere on the English countryside, and pursue his hobbies of shooting and catching rats.

## DARWIN'S CRUISE

He arrived home one day to find a letter from Captain Fitzroy, a highly respected sea captain who commanded the **Beagle**. Charles was invited to take the position of naturalist for a five year ocean voyage that would take him to exotic locations. Darwin decided that sailing the ocean was more exciting than sipping tea on little old ladies' front porches in some obscure village. And the pay was much better than that of a country parson.

Many years after his return from that round-the-world ocean voyage, a voyage that took him from belief to unbelief, Darwin began to develop his philosophy of origins which resulted in the shaking of the philosophical and scientific foundations of the western world and secularization of society. He capitalized on man's innate tendency to strive for independence from God, and he snapped any chain that bound them together. He set man adrift in a world without purpose and without hope.

He did so by stealing ideas from his grandfather (whom he never credited) and from Aristotle. He often referred to "my" theory of evolution, when educated people knew that Aristotle, Anaximander and others had suggested, hundreds of years before Christ, that man came from animals. In fact, Darwin was forced to correct himself in the third edition of **Origin** in 1861. There he admitted that others had preceded him with the suggestion that the species have changed from lower to higher forms. So much for "his" theory of evolution!

Darwin had removed God from the picture, leaving men completely free to believe and live as they pleased! **That** is at the very bottom of the evolution-creation dispute. Evolutionists can never accept a personal God who created man because then they would be required to give a personal accounting to that God. After all, God reminds us in Hebrews 9:27, "And it is appointed unto men once to die, but after this the judgment." Of course, judgment is coming whether men recognize it or not!

It is said that Darwin developed his evolutionary theory while on the **Beagle**, but that is not correct. He developed some doubts on the voyage, but his theory came much later. He still believed the Bible during the voyage, and believed in the fixity of species until he returned to England.

Darwin admitted, "I did not then in the least doubt the strict and literal truth of every word of the Bible...."[2] Some have questioned whether Darwin really believed, but surely we must take his word for it. Of course, I am not suggesting that he had accepted Christ as personal Savior. Darwin's disciples depict him debating his evolutionary ideas with Captain Fitzroy (an outspoken Christian) during the long voyage, but that is another myth.

Charles went even further in his confession by saying, "Whilst on board the Beagle I was quite orthodox, and I remember being heartily laughed at by several of the officers (though themselves orthodox) for quoting the Bible as an unanswerable authority on some point of morality."[3] So Darwin admitted that he endured some ridicule for his belief in the orthodox position!

However, Darwin began to move away from biblical truth as he observed nature, and made copious notes of his observations. He noticed the 13 varieties of finches and decided that the facts conflicted with the teaching of Genesis; there was, however, no conflict with Genesis but with the accepted interpretation of Genesis! There is a difference, you know.

Most theologians of that day provided no room for change within a species, and that, as we know, is too rigid a view and not supported by science or the Bible. Of course, creatures change (within a species) but never change from one species to another--macroevolution.

Before sailing on the **Beagle**, a preacher-friend gave Darwin the first volume of Lyell's **Principles of Geology** with the warning that he should not believe it! Lyell, a lawyer, is credited with being the main proponent of the doctrine of uniformitarianism which basically holds that past geological processes operated in the same way and at the same rate as they do today. (That would eliminate the Genesis Flood.) Darwin's experience of being influenced by Lyell is a good reason for all of us to be careful and critical of everything we read.

## DARWIN'S CONFUSION

As Darwin got older, he became less convinced of his views, as admitted by scientist Michael Denton, "The popular conception of a triumphant Darwin increasingly confident after 1859 in his views of evolution is a travesty. On the contrary, by the time the last edition of the **Origin** was published in 1872, he had become plagued with self-doubt and frustrated by his inability to meet the many objections which had been leveled at his theory."[4] Now, since Darwin had doubts as to gradualism, don't you think everyone (including scientists who burn incense to Darwin) should ask, "But what if Darwin was wrong?"

Many have assumed that Darwin's opponents were motivated in their opposition by religious convictions; however, Denton concludes that, "it was the absence of factual evidence which was the primary source of their scepticism and not religious prejudice."[5]

Leading evolutionist Stephen Gould concurs with Denton: "Contrary to popular belief, no serious nineteenth-century scientist--not even the most theological catastrophist--argued for the direct intervention of God in the earth's affairs."[6]

Well, of course, that is a huge exaggeration, for how could there be a creation without the intervention of the Creator? Obviously Gould was not in the most lucid state when he wrote that statement, but his implication is correct in that Darwin's opposition were not motivated by religion, but by their scientific convictions. (Of course, what one believes about God does influence what he believes about other things.) It is a fact that the creationist-scientists of his day gave Darwin some reason to question his basic positions.

Loren Eiseley provides further proof of Darwin's lack of confidence: "A close examination of the last edition of the **Origin** reveals that in attempting on scattered pages to meet the objections being launched against his theory the much-laboured-upon volume had become contradictory....The last repairs to the **Origin** reveal...how very shaky Darwin's theoretical structure had become. His gracious ability to compromise had produced some striking inconsistencies. His book was already a classic, however, and these deviations for the most part passed unnoticed even by his enemies."[7] That is not the Charles Darwin most everyone knows. Note that his book has "striking inconsistencies."

Scientists like to present themselves as being objective, fair and above the common failures, faults and foibles of most people; however, that is not true. Darwin is a perfect example of unfairness and bias. One example of his bias and ineptitude is in his treatment of some South American Indians. In his **Voyage of the Beagle**, Darwin wrote of the Indians of Tierra del Fuego whom he described as being at a very low rung of the evolutionary ladder. He called them "the miserable inhabitants of Tierra del Fuego," and  characterized  them as being benighted cannibals, almost beasts.

However, Roman Catholic scholar and author, Paul Kildare wrote: "Darwin hardly saw an Indian at all, and could not speak one word of their language, yet his description of the Fuegians is still quoted as authoritative over a century later in countless so-called scientific works."[8]

Isn't it incredible that the scientists of the world would take the scribbling of a theology major as a scientific fact without any substantial evidence? It shows how quickly unbelieving scientists are to believe anything that would appear to destroy the truth of Scripture and support spurious scientific speculation. (World famous anthropologist Margaret Mead did a similar job on the Samoan people.)

Two missionary priests, both highly qualified scientists, who were on the staff of American and European universities, asserted that Darwin had no scientific qualifications at all. They further said: "The Fuegian Indians were not cannibals; they believed in one Supreme Being, to whom they prayed with confidence; they had 'high principles of morality' and they rightly regarded the white people who exploited them as morally inferior to themselves."[9] Well, well, Darwin was wrong again!

## RACISTS

Darwin and his disciples were not only unbelieving, pseudo-scientists, but they were also racists! Ernst Haeckel was a qualified biologist in Germany and a contemporary of Darwin who laid the foundation of racism and imperialism that resulted in Hitler's racist regime.

Adolph Hitler "...stressed and singled out the idea of

biological evolution as the most forceful weapon against traditional religion and he repeatedly condemned Christianity for its opposition to the teachings of evolution....For Hitler, evolution was the hallmark of modern science and culture, and he defended its veracity as tenaciously as Haeckel."[10]

Edward Simon, a Jewish biology professor at Purdue University, wrote: "I don't claim that Darwin and his theory of evolution brought on the holocaust; but I cannot deny that the theory of evolution, and the atheism it engendered, led to the moral climate that made a holocaust possible."[11]

I wonder what that "climate" is doing to students in the public schools as they are taught they came from animals and are without any purpose to life? Could the incredible number and depth of our social problems be the result of Darwinism? I am convinced this is so, for if one believes that life has no purpose, and man came from beasts, then dignity, fairness, kindness, honesty, faithfulness and justice have no relevance.

Sir Arthur Keith, a well-known evolutionist, assessed Darwin's impact on Hitler and Germany: "We see Hitler devoutly convinced that evolution produces the only real basis for a national policy....The means he adopted to secure the destiny of his race and people were organized slaughter, which has drenched Europe in blood...."[12]

The unreasonable, unbiblical, unscientific philosophy of Darwin and his disciples laid a foundation for hundreds of years of hatred, barbarity and unbelief reaching into the future and impacting millions of innocent lives.

If Darwin were alive today, he would be hooted out of the politically correct scientific community because of his outrageous views about Blacks. He thought they were closer to man's ape "ancestors" than the white race! Was Darwin a white supremacist? His statements indicate that fact.

Note Darwin's following opinion on race that evolutionists usually refuse to quote: "At some future period, not very distant as measured by centuries, the civilized races of man will almost certainly exterminate, and replace, the savage races throughout the world. At the same time, the anthropomorphous apes...will no doubt be exterminated. The break between man and his nearest allies will then be wider, for it will intervene between man in a more civilized state, as we may hope, even than the Caucasian, and some ape as low as a baboon, instead of as now between the negro (sic) or Australian and the gorilla."[13] Okay, Jesse, what do you think of Darwin, now?

Most people will agree that Darwin's major disciple, Thomas Huxley was a bigot when he compared blacks unfavorably to white men: "It may be quite true that some negroes are better than some white men, but no rational man, cognizant of the facts, believes that the average negro is the equal, still less the superior, of the average white man....The highest places in the hierarchy of civilization will assuredly not be within the reach of our dusky cousins...."[14]

"Dusky cousins!" Can you imagine how that would be received down at the National Association for the Advancement of Colored People? But it gets worse. Henry F. Osborne was Professor of biology and zoology at Columbia University and President of the American Museum of Natural History Board of Trustees wrote, "The Negroid stock is even more ancient than the Caucasian and Mongolians as may be proved by an examination not only of the brain, of the hair, of the bodily characteristics...but of the instincts, the intelligence. The standard of intelligence of the average adult Negro is similar to that of the eleven-year-old youth of the species Homo sapiens."[15]  Hey, some of the most bigoted red-necks don't believe that.

Another famous evolutionist who influenced Hitler was Edwin Conklin, Professor of Biology at Princeton University from 1908 to 1933 and President of the American Association for the Advancement of Science in 1936. That was the year prior to Hitler's Olympics. Conklin wrote, "Comparison of any modern race with the Neanderthal or Heidelberg types shows that all have changed, but probably the Negroid races more closely resemble the original stock than the white or yellow races.[16] So blacks are closer to alleged ape men than the other races! They haven't "evolved" as far and are closer to "the original stock."

But Ed went further and eliminated the possibility of ever becoming a member of the NAACP when he wrote, "Every consideration should lead those who believe in  the superiority of the white race to strive to preserve its purity and to establish and maintain the segregation of the races, for the longer this is maintained, the greater the preponderance of the white race will be."[17]

The major haters of the last 100 years have been evolutionists. Men like Nietzsche (who repeatedly asserted that God was dead, called for the breeding of a master race and for the annihilation of millions of misfits), Hitler, Mussolini, Marx, Engels, and Stalin have all been outspoken evolutionists, and these people and their theories have been responsible for the slaughter of multi-millions of people. It's amazing that so many liberals, radicals, fascists, communists, and the easily impressed worship at Darwin's shrine.

## DARWIN SAVED?

Many Christians have handed out literature, written by one of the most respected clergymen in North America, that tells of Darwin's alleged conversion through the efforts of Lady Hope. She is supposed to have visited him in his garden and found him reading the book of Hebrews. He supposedly told her that he was sorry for his evolutionary views, and that they represented the immature views of a youth. He, according to the story, accepted Christ as Savior.

Lady Hope reportedly gave her testimony at D. L. Moody's school in Northfield, Massachusetts. Following is the report by Lady Hope of Darwin's alleged conversion; you will have to decide for yourself if it is true:

He (Darwin) waved his hand toward the window as he pointed out the scene beyond, while in the other he held an open Bible, which he was always studying.

"What are you reading now?" I asked, as I seated myself at his bedside.

"Hebrews!" He answered, "still Hebrews, the Royal Book, I call it. Isn't it grand?"

Then placing his finger on certain passages, he commented on them.

I made some allusions to the strong opinions expressed by many persons on the history of the Creation, its grandeur, and then their treatment of the early chapters of the Book of Genesis.

He seemed greatly distressed, his fingers twitched nervously, and a look of agony came over his face as he said: "I was a young man with unformed ideas. I threw out queries, suggestions, wondering all the time over everything, and to my astonishment the ideas took like wildfire. People made a religion of them."

Then he paused, and after a few more sentences on the holiness of God and the grandeur of the Book, looking at the Bible which he was holding tenderly all the time, he suddenly said, "I have a summer-house in the garden, which holds about thirty people. It is over there," pointing through the window. "I want you very much to speak there. I know you read the Bible in the villages. Tomorrow afternoon I should like the servants on the place, some tenants and a few of the neighbours, to gather there. Will you speak to them?"

"What shall I speak about?" I asked.

"Christ Jesus," he replied in a clear emphatic voice, adding in a lower tone, "and His salvation. Is not that the past theme? And then I want you to sing some hymns with them. You lead on your small instrument, do you not?"

The wonderful look of brightness and animation on his face as he said this I shall never forget, for he added, "If you take the meeting at three o'clock this window will be open, and you will know that I am joining in with the singing."

How I wished that I could have made a picture of the fine old man and his beautiful surroundings on that memorable day.[18] So that is her story.

Evolutionists have vehemently rejected this report as have some creationists, but it is not an open and shut case either way. M. Bowden, in his excellent book, **The Rise of the Evolution Fraud**, has done more research than anyone, to my knowledge, on this issue.

Yes, Darwin fired a shot heard around the world, but Charlie fired a blank! It shook up the world, and gave support to unbelieving men who were determined that God did not create the world. But Darwin was not present at the creation; God was, and I'll take His word for what really occurred in the beginning rather than the word of an apostate preacher whose cockamamie theory is falling to pieces, and is now rejected by honest, informed, unbiased people.

Charles Darwin

## CHAPTER SEVEN

# EVOLUTION'S FOUNDATIONS ARE CRACKING

There are many theories concerning the arrival of man to his present state, but there are two basic possibilities: Either God created man as recorded in Genesis 1 and 2, or He did not. Charles Darwin and others have said that God had nothing to do with man's arrival at his present state. It all happened verrry slowly. However, let me remind you that Darwin was not present when God created everything. In fact, God did not even consult Darwin about His plans! Nor has anyone, anywhere, at anytime proved evolution to be even possible, much less true.

Evolutionists get all "aquiver" when they talk about how scientific they are, but informed people smile at them when they tout their "scientific" position. We smile because we know that evolution is as "scientific" as a voodoo-rooster-plucking ceremony in Haiti! Almost! Evolution is founded, not on science, but on distortions, myths, poor scholarship, circular reasoning, faulty premises, and a generous dose of wishful thinking. In plain English, it is pure quackery!

I repeat evolutionist George Gaylord Simpson who said, "It is inherent in any definition of science that statements that cannot be checked by observation are not really about anything...or at the least, they are not science."[1] Now that's an evolutionist speaking the truth.

Simpson was right. Evolution is called scientific; however, it does not meet scientific criteria: It is not observable, not repeatable, nor is it subject to experimentation. So whatever evolution is, it is **not** science. "Science" comes from the Latin word, **scio** meaning "to know." But evolutionists cannot "know." They think (sometimes), they believe, they hope, they assume, they suppose, etc. but they don't "know." They do make much of the "scientific method" while they **butcher** that method!

G. A. Kerkut, professor of Biochemistry wrote in his **Implications of Evolution**, "There are seven basic assumptions not mentioned during discussions of evolution...The first point I should like to make is that the seven assumptions by their nature are not capable of experimental verification...The evidence that supports (the general theory of evolution) is not sufficiently strong to allow us to consider it as anything more than a working hypothesis."[2] What's an hypothesis? Well, **Webster** says it is "an unproven theory." Some might call it "an educated guess."

John T. Bonner is an eminent biologist who wrote a review of Kerkut's book. In his review he wrote, "This is a book with a disturbing message; it points to some unseemly cracks in the foundations. One is disturbed because what it said gives us the uneasy feeling that we knew it for a long time deep down but were never willing to admit this even to ourselves..."[3]

Following are some good examples of the brutal attack upon the "scientific method" by evolutionists: In

**Origin**, and **Descent of Man**, Darwin tries to prove man's ascent from a one-cell creature to a banker in a three-piece suit by using the phrase "we may well suppose" over 800 times. Scientific? I think not.

H. G. Wells is guilty of the same unscientific folly in his **Outline of History** when he deals with man's alleged development from the swamps to the suburbs. Wells uses the followng phrases to connect one species to another: "is probably" or "was probably" 20 times; "it must have been" 12 times; "it would seem" 11 times; "it may have been" 9 times; "may or may not" 9 times; "perhaps" 5 times; "it seems to be" 5 times; "it is probable" 4 times; "possible" 3 times and "we may guess" 3 times. Very scientific, huh?

Wells continued to "con" us by writing such nonsense as "So far as we can guess"; "this is pure guess, of course"; "it is supposed"; "they suppose"; "if we assume"; "it appears to be"; "it is possible"; "whole story is fogged" (How about that one!); "confessedly jumbled"; and "inextricable mixed up"!

That last one is very appropriate for evolutionists. Since when do guesses take precedence over facts? That happens when you don't have any facts!

Is it any wonder that the foundation of evolution is cracking on all sides when some of their best promoters of the fairy tale insult the mind of a 12-year old? Yes, there are some major problems among evolutionists, but they usually try to give the impression of solidarity. Even Darwin expressed less confidence in his theory than modern "believers." Of course, many of those "believers" are scientists and college professors who have a vested interest in conning the public, but not all scientists have been conned.

## SCIENTISTS SPEAK

A good example of rumblings in the ranks of major scientists is revealed   by   world   famous   evolutionist, Theodosius Dobzhansky whose review of Pierre P. Grasse's **L'Evolution du Vivant** made waves around the world, big waves.

Dobzhansky wrote, "The book of Pierre P. Grasse is a **frontal attack on all kinds of 'Darwinism.'** Its purpose is 'to destroy the myth of evolution as a simply, understood, and explained  phenomenon,' and to show that evolution is a **mystery** about which little is, and perhaps can be known. Now, one can disagree with Grasse but not ignore him. He is **the most distinguished of French zoologists**, the editor of the 28 volumes of 'Traite de Zoologie,' author of numerous original investigations, and ex-president of the Academie des Sciences. His knowledge of the living world is **encyclopedic**, and his book is replete with interesting facts that any biologist would profit by knowing. Unfortunately, the theoretical interpretations of these facts leaves [sic] one dissatisfied,   and   occasionally   exasperated."[4]   (Emphasis added.) That astounding praise of Grasse (who no longer believes in Darwinism) is from an evolutionist!

Then Dobzhansky closes his review by writing, "The sentence with which Grasse ends his book is disturbing: 'It is possible that in this domain biology, impotent, yields the floor to metaphysics.'"[5] (God and special creation!) So, the critics of evolution are not just a handful of Ph.Ds and their snake handling fellow-travelers!

Grasse even calls Darwinism a "pseudoscience" in his, **The Evolution of Living Organisms**: "Present-day ultra-Darwinism, which is so sure of itself, impresses incompletely informed biologists, **misleads them, and inspires fallacious interpretations**....Through use and abuse of hidden postulates, of bold, often ill-founded extrapolations, a **pseudoscience has been created**. It is taking root in the very heart of biology and is **leading astray** many biochemists and biologists, who sincerely believe that the accuracy of fundamental concepts has been demonstrated, **which is not the case.**"[6] (Emphasis added.)

Paul Lemoine, director of the Natural History Museum in Paris and the editor of the **Encyclopedie Francaise** wrote: "The theories of evolution, with which our studious youth **have been deceived**, constitute actually a dogma that all the world continues to teach: but each, in his specialty, the zoologist or the botanist, ascertains that **none of the explanations furnished is adequate**....[I]t results from this summary, that the theory of **evolution is impossible**."[7] (Emphasis added.) Wonder why you've never heard that on a TV science special hosted by Carl Sagan?

Soren Lovtrup, Professor of Zoo-physiology at the University of Umea in Sweden trashed Darwin in his, **Darwinism: The Refutation of a Myth**: "...only one possibility remains: **the Darwinian theory** of natural selection, whether or not coupled with Mendelism, **is false**."[8] (Emphasis added.) And Lovtrup is no fundamentalist, either!

Scientist Stephen Stanley admits that gradualism did not happen: "The known fossil record fails to document a single example of phyletic evolution accomplishing a major morphologic transition and hence offers no evidence that the gradualism model can be valid."[9] Note that there is not one example!

Another highly respected scientist, giving devastating body blows to Darwin is  Lipson, a British physicist and Fellow of the Royal Society. He wrote, "I have always been slightly suspicious of the theory of evolution because of its ability to account for any property of living beings (the long neck of the giraffe, for example). I have therefore tried to see whether biological discoveries over the last thirty years or so fit in with Darwin's theory. I do not think that they do....To my mind, **the theory does not stand up at all**."[10] (Emphasis added.) It doesn't stand up, but it's being propped up!

Scientist and author, Michael Denton, responding to ardent disciples of Darwin stating evolution to be a fact, said, "Now of course **such claims are simply nonsense**. For Darwin's model of evolution is still very much a theory and still very much in doubt when it comes to macroevolutionary phenomena."[11] (Emphasis added.)

Dr. Edwin Conklin, biologist at Princeton University wrote in the January, 1963 edition of **Readers Digest**, "The probability of life originating from accident is comparable to the probability of the unabridged dictionary resulting from an explosion in a printing shop."[12]

Then Leavitt, President of Lehigh University said, "Protoplasm evolving a universe is a superstition more pitiable than paganism."[13]

Dr. T. N. Tahmisian, physiologist with the Atomic Energy Commission, said, "Scientists who go about teaching that evolution is a fact are great con men, and the story they are telling may be the greatest hoax ever. In explaining evolution **we do not have one iota of fact**."[14] (Emphasis added.)

Dr. Louis Bounoure, Director of Research at the

National Center of Scientific Research in France said, "Evolution is a fairy tale for grownups. This theory has helped nothing in the progress of science. It is useless."[15]

Now comes Heribert Nilsson, Director of Botany Institute, Lund University who admitted, "My attempts to demonstrate evolution by an experiment carried on for more than 40 years, have completely failed...It is not even possible to make a caricature of an evolution out of paleo-biological facts....The idea of an evolution rests on pure belief!"[16]

Dr. Albert Fleischmann, Professor of Zoology at the University of Erlangen in Germany said, "The Darwinian theory of descent has not a **single fact** to confirm it in the realm of nature. It is not the result of scientific research, but purely the product of imagination."[17] (Emphasis added.) Well well, a noted zoologist says evolution is "imagination."

Dr. Etheridge was the examiner at the British Museum who said, "Nine-tenths of the talk of evolutionists is sheer nonsense, not founded on observation and wholly unsupported by facts. This museum is full of proofs of the utter falsity of their views. In all this great museum, there is not a particle of evidence of the transmutation of species."[18]

Science writer, Aime Michel interviewed various French scientists, and his conclusions sent shock waves throughout the evolution fraternity: "The classical theory of evolution in its strict sense belongs to the past. Even if they do not publicly take a definite stand, **almost all** French specialists hold today strong mental reservations as to the validity of natural selection."[19] (Emphasis added.) Can you imagine what that statement does to arrogant evolutionists? No wonder they are running scared!

Other well-known scientists have boldly asserted that

evolution is "dead"; is a "myth"; has "no support in the fossil record"; has been known to be wrong by paleontologists "privately...for over a hundred years"; has had its "obituary" written; is "incoherent."[20] Yet Darwin has plenty of disciples who burn incense to him daily. They have become the deaf, dumb and blind priests of Evolutionary Religion.

## DARWIN'S DISCIPLES

Darwin has many fanatics who are dedicated to the humbuggery of evolution like F. B. Vance and D. F. Miller. They assured high school students that, "All reputable biologists have agreed that evolution of life on earth is an established fact,"[21] while J. M. Savage wrote in his college biology text, "No serious biologist today doubts the fact of evolution...We do not need a listing of evidences to demonstrate the fact of evolution any more than we need to demonstrate the existence of mountain ranges."[22] Are these men simply fanatics or fools? Maybe both? One thing is sure: they are bluffing their way through life.

Richard Goldschmidt, professor at the University of California before his death, stated: "Evolution of the animal and plant world is considered by all those entitled to judgment to be fact for which no further proof is needed."[23] Well, Dick surely didn't express any doubts about evolution, but of course, he had never seen evolution in progress nor did he ever see a transitional fossil.

Sir Julian Huxley showed far more confidence in the babblings of Darwin than did his grandfather, who was Darwin's "bulldog." Huxley pontificated: "The first point to make about Darwin's theory of evolution is that it is no longer a theory, but a fact. No serious scientist would deny the fact that evolution has occurred, just as he would not deny the fact that the Earth goes around the Sun."[24] To think they killed a tree to print such tripe!

What a dummy! Huxley, with the wave of his hand (amazingly created by God), dismisses many hundreds of scientists who believe in creation. I get the impression that Huxley is trying to convince **himself** that Darwin didn't fire a blank.

M. J. Kenny stated, "Of the fact of organic evolution there can at the present day be no reasonable doubt; the evidences for it are so overwhelming that those who reject it can only be the victims of ignorance or of prejudice."[25] Kenny speaks eloquent gibberish--pure psychobabble and has shown that the ignorant and the prejudicial are people like himself, who professing themselves to be wise have shown themselves to be fools.

Fools like Kenny, Sagan, Gould and Co. will never be convinced of creation because they are religious zealots, and have painted themselves into a very uncomfortable corner, while thinking people know the **Darwin** is sinking and are jumping ship. As we increase our knowledge of the operation of the universe, and the complexity of God's creation, more and more people are turning to the Designer for answers.

Grand Canyon Strata Formed as Result of World Flood

# CHAPTER EIGHT

# DESIGN PROVES CREATION

One of America's leading space scientists, Dr. Wernher van Braun, affirmed that the universe demands a Designer when he wrote, "One cannot be exposed to the law and order of the universe without concluding that there must be design and purpose behind it all....The better we understand the intricacies of the universe and all it harbors, the more reason we have found to marvel at the inherent design upon which it is based...."[1]

Dr. von Braun made two important points regarding the universe. One was that there was design and the other that there was purpose. These are two concepts that evolutionists wax eloquent in denouncing--unsuccessfully!

## SIMILARITIES

A person looks closely at a Ford automobile, then looks at a Lincoln. His conclusion: The Lincoln evolved from the Ford over a long process of time! Such is the non-reasoning of evolutionists. Just because two cars are similar, does not mean a gradual development took place. It means both cars had the same designer. Just as there are similarities between men and apes, it does not mean that men evolved from apes. It means that apes and men had the same

Designer--God!

Similarity does not equal relationship. Evolutionists often use similarity between animals and man to "prove" Darwinism. They point to the legs, neck, ears, etc. of apes and remind us how similar they are to those of men. Creationists likewise use similarities to support creation. We maintain that it would not be a wise God who developed a method of locomotion, hearing, eating, etc. then threw away His blueprint! God doesn't do dumb things. People do. So God used His blueprint for many of His creatures. Similarities don't mean common ancestry but a common Architect!

Only fools would say, "The automobile, with all its complex machinery, electronic equipment, computers, lights, etc. happened by chance." No, it had to be designed. Someone had to have an idea for an auto, then the idea had to be put on paper. Models were made and tested. Some ideas had to be changed, discarded or modified. Finally, after the various pieces of machinery were designed and produced, the fenders, bumpers, motor, side panels, grill, trunk, hood, doors, etc. were formed and put together on an assembly line, and finally an auto rolled off the line, ready to be driven.

Any person who suggests that the auto happened by chance over long periods of time from isolated iron ore, cloth, glass, etc. would be shipped out to the funny farm and receive state certification as a Qualified Nut. **That** is analogous to the evolutionists who scream hysterically that God did not create the universe, the world, and man. "It all happened gradually by accident," shouts the wild-eyed scientist from Hockumville as he struggles with the straps of his straight-jacket!

Isn't it incredible that bright men would have us believe that an auto (or the universe) simply happened? Of course, men will believe many outrageous things--as long as they are not found in the Bible!

It is easy to "see" what one **wants** to see if one does not look at all the facts in a logical way. If you observe robins nesting, you will note that their eggs are always blue; however, it is not logical (nor true) to conclude that any nest filled with blue eggs belongs to a robin. Darwin made that kind of mistake. He looked at what was apparent, saw what he wanted to see, and jumped to unsound, unscientific and untrue conclusions.

Darwin taught that design was only apparent, having been the result of purely "natural causes," always natural causes. There's never room for God!

Sir Julian Huxley said, "Darwinism removed the whole idea of God as the creator of organisms from the sphere of rational discussion. Darwin pointed out that no supernatural designer was needed; since natural selection could account for any known form of life, there was no room for a supernatural agency in its evolution."[2]

## DESIGN AND ORDER

Everyone admits the effect--we're here. However, a "cause" is not necessary, according to Darwin and his devotees. But is there not a cause behind every effect? Even the skeptic, David Hume said, "I never asserted so absurd a proposition as that anything might rise without a cause."[3]

Astronomer Robert Jastrow said it well: "Now we see how the astronomical evidence leads to a biblical view of the origin of the world. The details differ, but the essential elements in the astronomical and biblical accounts of Genesis are the same: the chain of events leading to man commenced suddenly and sharply at a definite moment in time, in a flash of light and energy."[4] One of our top astronomers seems to believe in God's design!

Jastrow also wrote that many scientists were surprised "by the discovery that the world had a beginning under conditions in which the known laws of physics are not valid, and as a product of forces or circumstances we cannot discover."[5] But he went further saying, "that there are what I or anyone would call **supernatural forces** at work is now, I think, a **scientifically proven fact**."[6] (Emphasis added.)

You can see God at work in His creation as reported by Cressy Morrison, former president of the New York Academy of Sciences. He suggests that it is obvious that the universe was erected by a great engineering Intelligence because we are exactly the right distance from the sun. If we were closer, we would be toasted, and if we were at a greater distance, we would freeze to death. If the Earth rotated one tenth as fast as it does, the days and nights would be ten times longer resulting in the sun burning up all the vegetation, and at night, the temperatures would freeze any remaining plants! If the moon were 50,000 miles from us, the ocean tides would submerge all the continents twice a day.[7]

Professor Harlow Shapley, director of Harvard Observatory, said, "I cannot imagine the planets getting together and deciding under what law they would operate. Nor do we find anywhere in the solar or stellar systems the debris that would necessarily accumulate if the universe had been operating at random."[8]

Design and order are the functions of the sun as its energy radiates upon Earth producing the climate and the storms on the sea and land.  It lifts billions of tons of water from the oceans, vaporizes and demineralizes it.  Then it collects the vapors into clouds and maneuvers them thousands of miles by the energy of the wind (which is generated by the sun).  The vapors are condensed, dropping life-giving rain to Earth.  What a miracle!

The tiny acorn is activated by the sun, causing it to grow to produce a huge tree with the ability to reproduce itself ad infinitum.  In each seed is all the information that determines the leaf structure, color of bark, size and strength of limbs, etc.  The green leaves of trees and plants with miraculous chlorophyll would not be possible without the sun.  All designed by a loving Creator!

Have you wondered how the universe operates so smoothly without the components banging into each other if it is the result of an explosion?  An explosion produces debris not design and order.

Author A. Sippert said it well in his book, **From Eternity to Eternity**, "[God] must also design a way of keeping them apart as they speed around in space as a solar system, planet systems and moon systems.  To keep them in place two different forces had to be designed and created.  The one force is gravity that keeps them from flying off into space and into each other.  Then another force had to be created--the right amount of speed around the mother sphere to counteract or balance the force of gravity.  According to the evolutionary belief or creed all this was supposed to have happened by itself, by chance."[9]

When you look up at Mount Rushmore at the

massive faces of four of our presidents, you are awed at the world's largest sculpture. You think of Gutzon Borglum who had the idea, then made the sketches, planned the physical approach, gathered the required tools, scaffolding, etc. then produced the faces. Only a fool suggests that the four faces resulted from natural causes. No sane, sensible person suggests that erosion by wind, rain, frost, ice, etc. could have **ever** produced such an effect. We realize that the effect, that is, the faces had a cause, i.e., the sculptor who was the designer and producer.

## AMAZING!

God created an amazing world! There are cheetahs that can run 70 miles per hour; insects that sleep for 17 years; seals that can stay under water for 45 minutes; eight-armed, ink-shooting octopuses that can eat their own arms and grow new ones; archer-fish that can shoot water 15 feet into the air and hit a bug; and hawks that can swoop down at 150 miles per hour and snatch away their prey.[10]

When you realize how intricately engineered the human body is, you become even more convinced of God's design. The human heart works without maintenance or lubrication for over 75 years, and is without comparison in the engineering world! In other words, man can't even come close to the work of God.

The joints, muscles and bones all work together, smoother than a well-oiled machine. Then look at the body's communication, circulatory, and control systems, and you must stand in awe of the incredible mind of God who put it all together.

When a woman gets pregnant, she begins to develop a new human being, and that human has the capacity to form **other** humans. (Of course, the same is true of other creatures.) The body of humans and animals has the astounding ability to not only reproduce itself, but also to **repair itself**! But before the body can repair itself, it must **review** the injury (or illness), **report** on the problem, then **resolve** the problem before **repair** begins. What machine can do **anything** similar to that? What fool would suggest that no design was involved?

The human brain is an incredible organ, awesome in its design and   function. The human brain holds enough information that would, if written in English, fill about twenty million volumes! That's the number of volumes in any of the world's largest libraries! In one human brain![11] "All by accident," say the evolutionists!

The human brain has about ten thousand million nerve cells, with **each** nerve cell putting out up to one hundred thousand connecting fibers by which it makes contact with other nerve cells in the brain. The total number of connections in the human brain is about a thousand million million![12] The **Science News Letter** said that the brain of man "is different and immeasurably more complicated than anything else in the known universe."[13]

Robert C. Cook tells us in **Human Fertility: The Modern Dilemma,** that the complexity of the human brain and nervous system is difficult to imagine. He states that a building the size of the pentagon would hardly be large enough to house a computer with as many synapses as the human brain. And it would take all the power of Grand Coulee to operate it and all the water of the Columbia River to dissipate the heat produced by its operation and to keep it from exploding into flames. After all that, the commuter could not perform such common mental functions as imagination and intuition. (And don't ever expect to see computers that actually **think**. It won't happen. I'd be thrilled if evolutionists would think.)

## DESIGNER DEMANDED

It would seem, to a thinking person, that the world around us demands a Designer. Look at a living cell, each part of which is not "alive." The complete cell **is** alive but not each component! It needs each part to function as a cell. A good analogy is an airplane that is made up of all non-flying parts. The tail section cannot fly, nor can the wings nor fuselage fly alone. However, when the components of the plane are combined according to the design, the plane flies. Like the airplane, the cell had a Designer.

Even a single-celled organism is incredibly complex. Carl Sagan of Cornell wrote, "A living cell is a marvel of detailed and complex architecture....The information content of a simple cell has been estimated ...to be comparable to about a hundred million pages of the **Encyclopaedia Britannica**."[14]

Some of us have carelessly spoken of the amoeba as if it is a blob of jelly, but that is not so according to evolutionist Simpson of Harvard, "Above the level of the virus, the simplest fully living unit is almost incredibly complex. It has become commonplace to speak of evolution from amoeba to man, as if the amoeba were the simple beginning of the process. On the contrary, if, as must almost necessarily be true, life arose as a simple molecular system, the progression from this state to that of the amoeba is at least as great as from amoeba to man."[15] What a statement, yet he is still a disciple of Darwin!

Darwin didn't know anything about genetics, and it's too bad. He might not have written his book had he acquired a working knowledge of genetics. Genetics alone, can refute evolution. How did the genetic code come into existence? It is associated with complex molecules known as DNA and RNA. Those molecules need a living cell to function; however, a cell, to function, must have the DNA and RNA genetic apparatus.[16] Which came first? You can't have one without the other!

The DNA is absolutely mind-boggling. If all the DNA in your body were placed end-to-end, it would stretch from the Earth to the moon over 500,000 times! And if all the coded information in your body were transferred to typewritten pages, it would fill the Grand Canyon fifty times![17]

But we are told that it all happened by chance! Yes, about like a cat, walking across piano keys and playing Beethoven's Unfinished Symphony!

## THE EYE

The human eye is another item that demands a Designer. Darwin, after the publication of his **Origin of Species**, expressed doubt to his friends as to "organs of extreme perfection" being developed by natural selection. He wrote to Asa Gray, an American biologist, two years after **Origins** was published, admitting, "The eye to this day gives me a cold shudder."[18]

He further commented on the eye saying, "To suppose that the eye, with all its inimitable contrivances for adjusting the focus to different distances, for admitting different amounts of light, and for the correction of spherical and chromatic aberration, could have been formed by natural selection, seems, I freely confess, absurd in the highest possible degree....The belief that an organ as perfect as the eye could have formed by natural selection is more than enough to stagger anyone."[19] But it didn't stagger him enough to make him discard his imaginative theory.

The eye is supposed to have evolved from a hole in the head to an organ far more sophisticated than the most modern camera! The human eye has automatic aiming, automatic focusing, and automatic aperture adjustment. It can function from almost total darkness to bright sunlight, see an object the diameter of a hair, and make about 100,000 separate motions in a day--all in living color! And while we sleep, it performs maintenance work![20]

The eye could not have evolved, because it would have been useless without its **complete** system of sight. It is a very complex organ whereby light is directed to the back of the eye on to super sensitive cells, and these send the information to the visual part of the brain. At that instant, you then "see" what you look at. But the eye without the cells would be useless, and the eye and the cells without the brain would be useless.

G. Hardin seemed to agree when he wrote, "How then are we to account for the evolution of such a complicated organ as the eye?...If even the slightest thing is wrong--if the retina is missing, or the lens opaque, or the dimensions in error--the eye fails to form a recognizable image and is consequently useless. Since it must be either perfect, or **perfectly useless**, how could it have evolved by small, successive, Darwinian steps?"[21] (Emphasis added.) Yes, indeed. Great question!

The answer is simple: the eye was formed at one time by an all wise Creator, not over millions of years.

Dr. David Menton, associate professor of anatomy at Washington University wrote, "Incredible structures such as the human eye make our most complicated computers look like beer-can openers."[22]

Evolutionists insist that the eye evolved over the ages, even though it would not have been useful until it had "matured" to perfection, and of course, there is not a **shred** of proof of its evolution. But it gets worse (for the evolutionist). The eye would have had to evolve **many** times for various creatures!

Professor Frank Salisbury dealt with this problem when he said: "My last doubt concerns so-called parallel evolution....Even something as complex as the eye has appeared several times; for example, in the squid, the vertebrates, and the arthropods. It's bad enough accounting for the origin of such things once, but the thought of producing them several times makes my head swim."[23] Mine too, Professor.

## INGENIOUS PLAN

Scientist Allan Sandage was overwhelmed by the ingenious plan of God in creation, as is obvious from his book: "...the world is too complicated in all its parts and interconnections to be due to chance alone. I am convinced that the existence of life with all its order in each of its organisms is simply too well put together. Each part of a living thing depends on all its other parts to function. How does each part know? How is each part specified at conception? The more one learns of biochemistry the more unbelievable it becomes unless there is some type of organizing principle--an architect."[24]

Throughout nature you can see God's design with the snowshoe rabbit being a good example. Each autumn in the northern hemisphere, the days get shorter. There's not as much sunshine because the northern half of the earth is tilted away from the direct rays of the sun. The snowshoe rabbit can tell the days are getting shorter and the nights longer, and the intelligent Creator causes her eyes to send that message to her brain. The rabbit's computer then flashes the signal to the pituitary gland in her head.

With the lack of sunshine going into the rabbit's eyes, her pituitary gland begins making special hormones. Only a few are made at first because the days aren't that short yet. The hormones travel into the bloodstream to the hair-producing cells in her body, and tell the hair cells to start making little white hairs. Now, the new white hairs push out the old brown hairs of summer. She sheds her old brown coat for a new white one for winter, making her almost invisible in the snow. The brown coat for the summer and fall seasons gives her protection during that time of year, and the white coat gives protection for the winter![25] All by design!

Now let's deal with migrating birds, specifically the whitethroated warbler that spends the summer in Germany, but winters at the head waters of the Nile River in Egypt!

Young birds are able to accomplish seemingly impossible tasks, unaided except by the instinct they received from the Master Designer.

The birds "...take off and fly, unguided, across thousands of miles of unfamiliar land and sea to join their parents, and they have never been there before. How do they navigate? German researchers raised some of the warblers entirely in a planetarium building. Experiments proved that in their little bird brains is the inherited knowledge of how to tell direction, latitude, and longitude by the stars, plus a calendar and a clock, plus the necessary navigational data to enable them to fly unguided to the precise place on the globe where they can join their parents!"[26] Where did the birds get the knowledge to accomplish such impossible tasks? It was by design.

Evolutionists can't explain how the golden plover travels about 8,000 miles south from the Hudson Bay region, crossing about 2,000 miles over the sea from Nova Scotia to the Caribbean nations and then spends the winter in Argentina!

The barn swallow migrates about 9,000 miles from northern Canada to Argentina, and the arctic tern migrates some 14,000 miles each year traveling from pole to pole and back!

Symbiosis indicates that God designed the Earth and creatures to interact with each other. The yucca plant can be fertilized **only** by the yucca moth (the moth being dependent on the plant all its life). Neither could live without the other! So which one "evolved" first?

One of the most incredible creatures in nature is the bombardier beetle. When it is under attack, it aims its two "tail guns" at the enemy. Vapors, heated to 212 degrees Fahrenheit and containing a noxious, irritating chemical project from the two tubes at the same time and a miniature explosion takes place. Spiders, frogs, ants, etc. head for high ground! This mechanism is highly sophisticated, consisting of two adjacent storage sacs, combustion chambers, gun-like swivel tubes and the instinct to utilize them. All by chance?

There are also chemicals (hydroquinones, hydrogen peroxide and enzymes) in the correct amounts and in the correct places. Of course, the nerve and muscle attachments must co-ordinate the whole operation and focus the protective spray. The spray is poisonous and malodorous, but the beetle never poisons itself![27]

How could such a creature evolve over millions of years without blowing off his rear end or poisoning itself? How could the beetle have survived long enough to develop its firing abilities without those protective abilities? Have you noticed how silent the evolutionists are about these incredible creatures?

The sonar of a bat is almost miraculous. My wife and I watched thousands of bats at the Carlsbad (NM) Caverns fly from the cave entrance without one flying into another. Since the bats are almost blind, God gave them a fantastic ability to "see" via built-in sonar.

A bat detects its own sound, as differentiated from thousands of others, even if the signal is 2,000 times fainter than background noises! It can "see" a fruit-fly up to 100 feet away by echo location and catch four or five in a second! And this amazing system weighs a fraction of a gram!

The bat's ability makes man's radar and sonar look like tin can operations. How did such a system evolve over millions of years? It had to work perfectly from the beginning or the bat could not have survived to develop it.

Central and South America is home to the Bull's Horn Acacia tree that has large hollow thorns inhabited by ferocious stinging ants. The tree supplies food and shelter to the ants and the ants protect the tree from animal predators and plant competitors. The ants are gardeners and eat every green shoot that grows near the tree providing space and sunlight in a tropical jungle where both are limited. When the ants are removed, the tree dies in 2-15 months.[28]  Again, all nature screams, "design."

Another example of God's creative design is fish that roam the ocean devouring smaller fish, only to find that their mouths are filling with parasites. The large fish opens its mouth, exposing its ferocious-looking teeth and a small cleaner fish swims into the mouth cavity and cleans the teeth and mouth **without being eaten**! When the cleaning job is finished, the small fish swims out of the larger fish's mouth unscathed (and fully satisfied) as the cleaned fish swims away. Everyone is happy! All by accident?

A similar incident takes place with the Egyptian plover (a bird) that walks over the lip of the Nile crocodile and into its mouth to clean out parasites! The crocodile never harms the tasty bird! "But," say the evolutionists, "it all happened by chance." Sure it did, and if you believe that, I have a bridge in San Francisco to sell you.

Creationists affirm that all these indications of design were part of God's plan for His glory and our good. The Psalmist exulted in Psalms 104:24, "O Lord how manifold are thy works. In wisdom hast thou made them all. The earth is full of thy riches."

But, of course, design is anathema if you are trying to hide from the Designer!

# CHAPTER NINE

# NO MISSING LINKS

If man has slowly evolved from lower forms of life to his present state without any design and direction from God, there would be millions of fossils to document that journey; however, that is not the case. Out of billions of fossils found in the past 100 years, not one is an in-between, half and half link! Not one. Even the **Encyclopaedia Britannica** admits, "Direct fossil evidence of the earliest members of the human species, **Homo sapiens**, is rather scarce."[1] "Rather scarce!" There is none!

Dr. Duane Gish is a well-known biochemist who stated, "Certainly it can be said without fear of contradiction that the evolutionary ancestors of the Cambrian fauna, if they ever existed, have never been found."[2]

Darwin repeatedly admitted this problem. He wrote, "...the distinctness of specific forms and their not being blended together by innumerable transitional links is a very obvious difficulty."[3]

It was thought, in Darwin's day, that the major problem of gaps would be solved with the advance of the science of paleontology (the study of fossils); however, that has not happened, as admitted by David Kitts, professor of geology at the University of Oklahoma.

Dr. Kitts wrote, "Despite the bright promise that

paleontology provides a means of 'seeing' evolution, it has presented some nasty difficulties for evolutionists, the most notorious of which is the presence of 'gaps' in the fossil record. Evolution requires intermediate forms between species and paleontology does not provide them...."[4]

David Raup, Professor of Geology at the University of Chicago, and a major evolutionist, referred to Darwin's hopes for future discoveries: "The evidence we find in the geologic record is not nearly as compatible with darwinian natural selection as we would like it to be. Darwin was completely aware of this. He was embarrassed by the fossil record because it didn't look the way he predicted it would and, as a result, he devoted a long section of his **Origin of Species** to an attempt to explain and rationalize the differences."[5]

Darwin expected future paleontology to support his position; however, the opposite has taken place! New evidence has required some of the "transitions" to be discarded as I shall discuss later in this chapter.

## GAPS

The gaps are real and all the wishing, looking, hoping and longing for them to be filled will not change that fact. The fossils prove evolution to be only a pitiful theory, and a third rate one at that.

T. N. George, writing in **Science Progress**, confessed that, "There is no need to apologize any longer for the poverty of the fossil record. In some ways it has become almost unmanageably rich, and discovery is outpacing integration...The fossil record nevertheless continues to be composed mostly of gaps."[6]

The world should be full of the bones of half-men, but they are not there; not there because they never lived! What a pity that evolutionists are looking for something that never existed! Not much chance for success! If Darwin's gradualism took place, how has it managed to escape detection?

Few critics of evolution note that there is a great scarcity of **any** kind of human remains. Forget the alleged ape-men, and consider **full** man who has been roaming the earth for millions of years (according to the fairy tale). Where are his bones? Only a few have been found. When an evolutionist is faced with this problem, he tells us that few bones have been found of early  man because they buried their dead early in history.

All right, let's accept that as true (because it is true), but that fact proves that early man respected the dead and recognized that man had a soul that would be resurrected. So their respect for the dead was made evident in their burial of the dead rather than permitting them to lie where they fell as do the beasts of the field.

Evolutionists have been looking for the links between various species, especially between apes and man for more than a hundred years without success. Their disappointments have been many. One was **Ramapithecus**, who held such promise, but now  is admitted to have been, not a link between ape and men, but only an ape! Eckhardt concluded the "entire **Ramapithecus**, walking upright, has been 'reconstructed' from only jaws and teeth. In 1961, an ancestral human was badly wanted. [**Ramapithecus**] latched onto the  position by his teeth, and has been hanging on ever since...."[7] Evolutionists get what they want even if they have to construct it from a handful of bones!

Dr. David Pilbeam of Yale University is a physical anthropologist with a national reputation, and he reluctantly rejected **Ramapithecus** in a revealing admission: "In the course of rethinking my ideas about human evolution, I have changed somewhat as a scientist. I am aware of the prevalence of implicit assumptions and try harder to dig them out of my own thinking. I am also aware that there are many assumptions I will get at only later, when today's thoughts turn into yesterday's misconceptions. I know that, at least in paleoanthropology, data are still so sparse that **theory heavily influences interpretations**. Theories have, in the past, clearly reflected our **current ideologies instead of the actual data**."[8] (Emphasis added.)

Wow! What an admission! Pilbeam is still an evolutionist, yet he admits what creationists have known for many years: the evolutionist theory heavily influences interpretations! And they surrender their theories reluctantly, and usually not in publications!

However, one of the world's best known paleontologists, Niles Eldredge, admitted that deception has been common. He wrote, "We paleontologists have said that the history of life supports [the story of gradual adaptive change], all the while really knowing that it does not."[9]

Lord Zuckerman is one of England's most respected scientists who has conducted many years of research on the skeletal structure of apes and men. His research led him to admit, "...[if man] evolved from some ape-like creature...[it was] without leaving any fossil traces of the steps of the transformation."[10] He admitted that there are no "fossil traces" of transformation from an ape-like creature to man. None!

Zuckerman further admitted, "...[in] the interpretation of man's fossil history, **where to the faithful anything is possible**...the ardent believer is sometimes able to believe **several contradictory things at the same time**."[11] (Emphasis added.)

One evolutionist, in a seizure of honesty, confessed, "It seems that every time a man finds a human bone he goes crazy on the spot."[12]

"All right," says the honest inquirer, "what about stone-age men? Surely they lived." Yes, they did. For thousands of years, stone-age people have lived in various parts of the Earth, and some are living today in places like the Philippine Islands.

Evolutionists are desperate to find missing links, so they find them where they don't exist! Their faith demands it! And they look for the links between all the species, without success. But they keep looking, hoping, and praying--no, not praying, unless there is something called, a "secular prayer." They seem to believe a Bible principle--seek and ye shall find. If they seek with enough prejudice, they will find **whatever** they are looking for.

Lord Zuckerman recognized this problem, and admitted that the field is overrun with wishful thinking and myths. He said, "The [fossil] record is so astonishing that it is legitimate to ask whether much science is yet to be found in this field at all. The study of the Piltdown Man provides a pretty good answer."[13] (Piltdown was a total fraud. See chapter eighteen.)

Evolutionists would rather cling to a fairy tale than

rest on a fact. The gaps **are** filled--with guesses. There is nothing--I repeat-- nothing to link man to the apes. And it's also true of other creatures.

## NO EVIDENCE

If fish evolved into amphibians (animals that live in sea and on land), as evolutionists teach, then there would be millions of fossils showing the gradual transition of fins into feet and legs. Of course, the anatomy and physiology of fish would have to be changed to accommodate the new environment of land. However, Darwin admitted that, "Not one change of species into another is on record...we cannot prove that a single species has been changed."[14] That's some admission from the Guru of Gradualism.

One of the world's greatest scientists, Louis Agassiz, of Harvard, told us that the links between the species are imaginary! He wrote, "Species **appear suddenly** and disappear suddenly in progressive strata. That is the fact proclaimed by Paleontology...The geological record, even with all its imperfections, exaggerated to distortion, tells now, what it has told from the beginning, that the **supposed intermediate forms** between the species of different geological periods are **imaginary** beings, called up merely in support of a fanciful theory."[15] (Emphasis added.)

That statement comes from a world-renowned scientist who was a creationist! Now, an admission from a confirmed evolutionist, Dr. Stephen J. Gould, Professor of Paleontology at Harvard University: "The fossil record with its abrupt transitions offers **no support** for gradual change...."[16] (Emphasis added.) You must remember that fossils are Gould's business, and he is considered one of the best. His conclusion: there are no fossils that prove a gradual change from one species to another!

Dr. Niles Eldredge is a supporter of Stephen Gould's punctuated equilibrium theory, and is quoted by the British newspaper, **The Guardian Weekly** in an article titled, "Missing Believed Non-existent." The reporter wrote, "If life had evolved into its wondrous profusion of creatures little by little, Dr. Eldredge argues, then one would expect to find fossils of transitional creatures which were a bit like what went before them and a bit like what came after. But no one has yet found any evidence of such transitional creatures. This oddity has been attributed to gaps in the fossil record which gradualists expected to fill when rock strata of the proper age had been found. In the last decade, however, geologists have found rock layers of all divisions of the last 500 million years and **no transitional forms were contained in them**."[17] (Emphasis added.)

Eldredge also said, "The smooth transition from one form of life to another which is implied in the theory is...not borne out by the facts. The search for 'missing links' between various living creatures, like humans and apes, is probably fruitless...because they probably never existed as distinct transitional types...no one has yet found any evidence of such transitional creatures."[18]  That's from another expert on fossils!

If reptiles evolved into birds then we would find intermediate fossil forms showing the gradual change of the forelimbs of the reptile into the wings of a bird and the gradual formation of feathers on the reptile. But those transitional fossils have not been found because they never existed.

This is admitted by evolutionist W. E. Swinton who stated, "The origin of birds is largely a matter of deduction. There is no fossil evidence of the stages through which the remarkable change from reptile to bird was achieved."[19]

## TRUE BIRD

Desperate evolutionists grasp for **Archaeopteryx** as their proof of a transitional form between reptiles and birds. Dr. Henry Morris replied in his classic, **The Biblical Basis for Modern Science**, "**Archaeopteryx** had teeth and claws like a reptile, wings and feathers like a bird, but all were perfectly formed for the needs of the creature. It did not have 'sceathers' (half-scales, half-feathers) or 'lings' (half-legs, half-wings)."[20] (By the way, what good would a half-limb, half-wing be anyway? It would be a helpless creature that could not survive to evolve into another creature.) Morris adds that **Archaeopteryx** could not have been the link between reptiles and birds because birds have existed as long as **Archaeopteryx**, so it could not have been the ancestor of birds.

Morris is supported by paleontologist James Jensen who found **modern** birds in the same rock strata in which **Archaeopteryx** was found, so **Archaeopteryx** lived at the same time as various other birds![21]

Sir Gavin de Beer evaluated scientific papers on **Archaeopteryx** with interesting results. Six experts called it a lizard, eight said it was a transitional creature between reptiles and birds, and thirty-seven classified it as a bird![22]

So what's the debate about this bird? Well, it seems

that it had feathers (like a bird) but teeth and claws (like a reptile). However, teeth and claws do not a reptile make. Some reptiles, like turtles, do not have teeth. But what about the claws? Well, the ostrich has claws, but no one calls it a reptile. The hoatzin of South America and the touraco of Africa both have small claws yet they are still birds--poor flyers, but still birds.[23]

Poor evolutionists! Their best fossil link proves to be simply another bird. In the eighty million years it allegedly took for reptiles to evolve into birds, only **one**, **Archaeopteryx**, has been found, and now it turns out to be a farce. In fact, some scientists have charged that the feathers on **Archaeopteryx** were forged! Whatever the facts as concerning the alleged forgery, Michael Denton and other evolutionists admit that it is a true bird, not a link between reptiles and birds.[24] Even T. H. Huxley agreed it was "all bird."[25]

## PHANTOMS

World famous paleontologist Colin Patterson said in a letter regarding links between species, "Gould and the American Museum people are hard to contradict when they say that there are **no transitional fossils.** As a paleontologist myself, I am much occupied with the philosophical problems of identifying ancestral forms in the fossil record. You say that I should at least 'show a photo of the fossil from which each type organism was derived.' I will lay it on the line-- **there is not one such fossil for which one could make a watertight argument.**"[26] (Emphasis added.) Note that he said, "not one." Zero, zilch, none!

Maybe that's the reason evolutionist Dr. Mark Ridley, who is a biologist at Oxford, states that "no real evolutionist, whether gradualist or punctuationist, uses the fossil record as evidence in favor of the theory of evolution as opposed to special creation."[27]

Dr. Colin Patterson admits that the textbooks are, **"replete with phantoms**--extinct, uncharacterizable groups giving rise one to another."[28] (Emphasis added.) One such phantom is the horse series that almost all textbooks use as one of the **best** examples of Darwinian gradualism.

Sir Gavin de Beer, Director of the British Museum of Natural History, in a speech to a group of geologists said of the horse series, 'Now this series is **incontrovertible**. It provides clear evidence of the transition of one genus into another over a period of something like seventy million years."[29] No, not quiet "incontrovertible." In fact, he was wrong, dead wrong!

Everyone has seen the horse series in museums and textbooks, but I'm sorry folks, you've been had! The horse series has been pulled from the museums and textbooks (at least those interested in accuracy). It turns out to be a phantom.

Students were never told that the seven horse fossils are never found in any one place in the world, nor are they told that the series starts in North America, then goes to Europe, then back to America! That takes some evolving!

Evolutionist G. G. Simpson, of Harvard University, wrote of the horse series, "The uniform continuous transformation of **Hyracotherium** into **Equus**, so dear to the hearts of generations of textbook writers, never happened in nature."[30]  Simpson also said the picture of an exhibit at the American Museum of Natural History is "flatly fictitious."[31]

Some experts don't think the smallest "horse" is even a horse! And when a horse can't be counted on to be a horse, then of course, you have trouble, real trouble.

Dr. David Raup, formerly of the Chicago's Field Museum of Natural History, and now Professor of Geology, University of Chicago, admits that the horse series has "had to be discarded or modified."[32] He also said, "we have to abandon belief in the evolution of the horse."[33]

Saiff and Macbeth very candidly admit that the horse series is "very deceptive." They write of "a **very deceptive picture** that has been presented over and over again in the literature. This is the famous chart that shows, in seven neatly graded steps, the supposed descent of the horses from tiny **Eohippus** to the modern **Equus**. This chart is so persuasive that most readers will be shocked to learn that it is an **illusion**.

"These seven stages **do not represent ancestors and descendants**. They are fossils that were taken from different times and places, and were then strung together, perhaps innocently, to show how evolution might have (or should have) handled the matter. The experts, good darwinians though they may be, do not contend that things actually occurred in this simple straightforward way. They regret that the chart was ever made and they **have tried to expunge it** from the record, but it persists despite their efforts and appears in one textbook after another..."[34] (Emphasis added.)

"Perhaps innocently!" You've got to be kidding. They took fossils from different continents, placed them in order according to size, even tinkered with the side toes to make them look more "horsey" and it was innocent! If you believe that, you also believe in Santa Claus and the Easter Bunny!

Desperate evolutionists are quick to believe anything that supports their myth, and often make inaccurate judgments based on spurious information. Note that they developed their notorious horse series from bones, bones found in different continents! (That takes **some** evolving.) Their final decisions are based on very limited facts, and when you add dishonesty to desperation, you usually get distortion.

Let me provide a good example of the above. If you mate a female horse with a male ass, the offspring is a sterile mule, but if we did not have that information and found them as fossils, they could be arranged in a series to **prove** how an ass evolved into the mule and the mule evolved into the horse. And that decision could be reached by an **honest** person looking for **proof** of evolution--and he would be **wrong** as he assembled his missing links.

Professor Heribert Nilsson of Lund University, Sweden has commented on the problem of missing links: "It is not even possible to make a caricature of evolution out of paleobiological facts. The fossil material is now so complete that the lack of transitional series cannot be explained by the scarcity of the material. The deficiencies are real, they will never be filled."[35] But don't tell high school teachers that fact.

It's a fact that there are no transitional fossils between the species. The only changes occur **within a group**. Of course creatures have changed since their creation, but never from lower to higher forms. Never!

Evolutionists taught for many years that a bony fish called the "coelacanth" that swam in the seas in abundance about "400 million years ago," was a perfect intermediate. They suggested that it lurched upon land searching for food, and gradually stayed longer until it adapted to the new environment. Then it disappeared about seventy million years ago--until 1938. Then one was pulled from the waters off the coast of Madagascar, exactly like the fossils of "400 million years ago!" No evolution! And with an embarrassing thud, coelacanth falls from the realm of transitional forms! What a shame! (About 100 other specimens have been pulled out of the water since then!)

Birds appear suddenly in the fossil record, as do mammals, with all their complexity. Lyall Watson acknowledged in **Science Digest** that primates appear abruptly in the fossil record. They "have sprung out of nowhere."[36] There are no simple creatures leading up to them so there is no evolution.

Finally, let's let Charles Darwin speak on the subject of the alleged gaps between the species: "...innumerable transitional forms must have existed but why do we not find them embedded in countless numbers in the crust of the earth?....why then is not every geological formation and every stratum full of such intermediate links? Geology assuredly does not reveal any such finely graduated organic chain; and this, perhaps, is the **most obvious and gravest objection which can be urged against my theory**."[37] (Emphasis added.)

I could go on and on for pages showing that evolutionists admit the absence of transitional fossils, but it is not necessary or profitable to overload you. I do want to point out again what scientists don't want to discuss, and most laymen don't know: the dishonesty and fraud among some, but not all, evolutionists.

Zoology Professor Ernst Haeckel of the University of Jena was an ardent supporter of Darwinian evolution, and he faked evidence to support his position. (Obviously, he wasn't very sure of his position!) He was caught and charged with fraud by five professors. When convicted by a university court at Jena, he admitted that he had "altered" his drawings of transitional fossils (to prove evolution).

It is to the credit of the university and the five professors that they called his bluff (or rather, his fraud); however, Haeckel was not dismissed, demoted or disgraced! His defense was incredible and enlightening: "A small per cent of my embryonic drawings are forgeries...[and] I should feel utterly condemned and annihilated by the admission, were it not that hundreds of the **best** observers and biologists lie under the same charge."[38] (Emphasis added.)

What an admission! Think of Haeckel when you visit museums and see the drawings and models of "missing links" that "prove" evolution. The "missing links" will always be missing because they never existed. Also remember his charge that "hundreds" of other scientists did exactly what he did! Of course, he should have been fired from the university, but obviously, he met the standard of mediocrity expected of many college professors!

Richard Leakey and Roger Lewin candidly admitted their dearth of information about one missing link, and the way they dealt with the lack of information. They wrote, "Now, if we are absolutely honest, we have to admit that we know nothing about **Ramapithecus**; we don't know what it looked like; we don't know what it did and naturally, we don't know how it did it. But with the aid of jaw and tooth fragments and one or two bits and pieces from arms and legs, all of which represent a couple of dozen individuals, we can make some **guesses**, more or less inspired."[39]

Yes sir, inspired guesses! Isn't "science" (meaning "knowledge") so reassuring?

# CHAPTER TEN

# WHAT ARE HOPEFUL MONSTERS?

As we have discussed in the previous chapter, one of the most devastating problems evolutionists have is the lack of transitional fossils or "missing links" to support their fairy tale. The missing links are still missing, so some brave, brash boys of science have done some non-thinking to circumvent the problem.

## THE PROBLEM

Dr. Stephen Gould, professor of Paleontology at Harvard University wrote, "The fossil record with its abrupt transitions offers no support for gradual change...."[1] In fact, Gould made an unusual admission at a meeting of evolutionists in Chicago in 1980. He told them that the absence of transitional fossils had been the "trade secret" of paleontologists for a long time.[2] However, he and others have not told sixth grade science teachers that fact. It is still a secret to many.

Darwin predicted that the fossil record, when it was more complete, would fill in the gaps. He was wrong--again. Top evolutionists, Eldredge and Tattersall confessed to Darwin's error when they wrote, "One hundred and twenty years of paleontological research later, it has become abundantly clear that the fossil record will not confirm this part of Darwin's predictions. Nor is the problem a miserably poor record. The fossil record simply shows this prediction was wrong."[3]

Harvard's world famous paleontologist, Louis Agassiz (a creationist to his death) confirmed the folly of evolution many years ago: "Species **appear suddenly** and disappear suddenly in progressive strata. That is the fact proclaimed by Paleontology....The geological record, even with all its imperfections, exaggerated to distortion, tells now, what it has told from the beginning, that the **supposed intermediate forms** between the species of different geological periods are **imaginary** beings, called up merely in support of a fanciful theory."[4]  What an indictment of evolution!

But the best witnesses against evolution are the evolutionists! Probably the main evolutionist in America is Stephen J. Gould who wrote, "New species almost always appeared suddenly in the fossil record with no intermediate links to ancestors in older rocks of the same region."[5]

Well, he makes it clear that there are **no** intermediate links from one species to another. That's what we creationists have been saying for years.

Another top evolutionist is G. G. Simpson, who said the same thing as Gould: "...it remains true, as every paleontologist knows, that most new species, genera, and families...appear in the record suddenly and are not led up to by known, gradual, completely continuous transitional sequences."[6]

So, what is the answer since the gradualism of Darwin is not the "fact" we have been taught? Will scientists now admit that in the beginning, God created everything? Afraid not. Scientists tell us, with a straight face, that, no, gradualism didn't happen, but evolution **happened all at once**! No, there aren't any gaps to fill because new species came about as a result of some fortuitous miracle. Of course, it isn't **called** a miracle, except by Darwin! (Yes, he thought the sudden appearance of a new organ would be a miracle.)

The highly respected scientist, Ernst Mayr wrote, "To believe that such a drastic mutation would produce a viable new type, capable of occupying a new adaptive zone, is equivalent to believing in miracles."[7]

As some critics of "sudden appearance" have said, this idea is simply a middle ground between creation and evolution. Saltation simply means that a new form popped up out of nowhere without anyone having any idea as to how, why or where it originated! I call it a "secular miracle."

The result of the "miracle" is new organs, and yes, even a different creature where the new and advanced species appears in a single leap! Let's go once more to the number one evolutionist in America, Stephen J. Gould: "There has been no steady progress in the higher development of organic design. We have had, instead, vast stretches of little or no change and one evolutionary burst that created the entire system."[8]

Well, of course, the first part of the statement is true, while the last part is false. He is trying to justify why there are no fossil links that prove evolution.

You can almost feel the desperation of these men. They now admit that there are no transitional fossils to link the species, so they have come up with the most "unheard of" theory I've ever heard of! There is no need for missing links because it all happened at once, so no fossils are there to be found by us today! We are told that the answer is in "hopeful monsters," however, they are really "hopeless monsters."

Stephen Gould wrote, "Macroevolution proceeds by the rare success of these hopeful monsters, not by continuous small changes within populations."[9]

The number one scientist in America just admitted that Darwin's evolution did **not** take place as is being taught in all the public schools in America! Gradualism is false!

Gould has adjusted his evolutionary theory to get around the problem of no scientific support for gradual development of molecules to man, and calls his theory (guess) "punctuated equilibrium."

Gould and compatriot Niles Eldridge tell us that evolution happened in short, sharp jerks that were followed by stable periods during which very little change took place. Of course there is **no** evidence for this absurd theory, but it does solve the problem of lack of fossil transitional forms between species, although it provides no answer whatsoever for the complete lack of transitional forms between higher categories, such as genera, orders, classes, and phyla.

These desperate punctuationalists have received much attention, but that attention has resulted in ridicule for their outrageous attempts to deny Genesis 1 and 2. Scientist and author Michael Denton admits that it would take "miracles" to bridge the gaps in the fossil record.

Denton comments on punctuated equilibrium, stating that it "...has been widely publicized but, ironically, while the theory was developed specifically to account for the absence of transitional varieties between species, its major effect seems to have been to draw widespread attention to the gaps in the fossil record."[10]

We creationists have been saying that the fossils do not prove evolution; they give support to creation. And now we are supported by the very ones we debated for years! Thanks, fellows, for substantiating what you have been denying for so long!

Stephen J. Gould is still saying that evolution took place; it just didn't take place gradually over millions of years. He has not become a creationist; in fact, he is a Marxist! Poor fellow has three strikes against him: He is an evolutionist, he is a Marxist (can you believe that, in our political climate?), and he is a Harvard professor. Too bad.

## MEET HOPEFUL MONSTERS

Gould has revived the "hopeful monster" theory that was birthed by paleontologist O. H. Schindewolf and geneticist Richard Goldschmidt in the 1930s and 1940s. Goldschmidt was a German Jew who served as Professor of Zoology at the University of California, Berkeley and taught that major categories arose instantaneously. He called this mechanism, the "hopeful monster" mechanism. The really amazing thing is that his fellow scientists didn't fall to the floor holding their sides with laughter, gasping for breath, when he presented his theory. Most of them only snickered; a few were converted.

Goldschmidt admitted, "The facts of greatest general importance are the following. When a new phylum, class, or order appears, there follows a quick, explosive (in terms of geological time) diversification so that practically all orders or families known appear suddenly and without any apparent transitions."[11]

Stephen Stanley, of Johns Hopkins, wrote of Goldschmidt's buddy, O. H. Schindewolf, "Schindewolf believed that a single cross mutation could instantaneously yield a form representing a new family or order of animals. This view engendered such visions as the first bird hatching from a reptile egg."[12] Wow, doesn't that scream, "desperation"?

A sensible person, without an anti-God bias, would have said, "The facts of geology, zoology, paleontology, etc. scream special creation as to the origin of the universe and man," however, Goldschmidt would not consider that possibility so he came up with his cockamamie "hopeful monster" theory.

Goldschmidt suggested that many millions of years ago, there was not a bird upon the face of the Earth; however, there were reptiles. No one can explain what happened (ergo they don't have to), but one day, a reptile got pregnant, and laid an egg. The egg hatched and out jumped a bird! (Frankly, I think Goldschmidt laid the egg, and it was rotten!)

He is telling us that we don't need to fill in the fossil gaps with transitional fossils, because there are no fossils of missing links since different species came into existence all at once--like the bird hatching from a reptile egg! My, my, what extremes unbelieving men will go to in order to exclude God from their lives. But, at least Goldschmidt forced his fellow scientists to admit the failure of Darwin's gradualism.

All right, let's play their game and deal with the absurd possibility of a bird hatching from a reptile's egg.

(Can you imagine what that did to the mother reptile when a bird jumped out?) Now, remember that this bird has hatched contrary to the basic law of biology: like begets like. God set the boundary in Genesis 1:24 when He commanded each creature to reproduce "after his kind." This is the boundary over which beasts and men cannot cross, but Goldschmidt's theory disregards that basic law of biology.

All right, we have **one** bird upon the face of the Earth. Big deal, what good does that do? It takes **two** to start a species, unless Goldschmidt has repealed **that** law as well. I can see this hopeful monster searching the countryside for another monster of the opposite sex. Let's also suppose it happened again, but the birds were the same sex. Or what if they were hatched thousands of miles away from each other?

There are other possible problems: What if the two birds are hatched, but one is a huge eagle and the other is a petite hummingbird? No way to reproduce. And what happens if one is a vicious falcon that feeds on small birds? Then it would be **meal** time not mating time!

I suppose if one is naive enough to believe a reptile could give birth to a bird, he could also be convinced that the reptile would give birth to **two** birds. And let's make them male and female, and let's suppose they both survive and begin to mate. What if they are not fertile? How do they know how to take care of themselves and their young? Big question: How do they learn to fly? After all, it isn't in their genes!

Of course no sane person would suggest that a reptile egg could hatch a bird, but don't bet the rent money on it! Sensible people are saying, "It is incredible that anyone could teach such an outrageous possibility with a straight face."

Yes, but look at this additional problem: That same "miracle" would have had to happen many times to produce **every** known species that ever existed on the Earth!

Goldschmidt, while a reputable scientist, did receive much deserved ridicule for his nonsensical theory, but Stephen Gould ventured to say: "I do, however, predict that during this decade Goldschmidt will be largely vindicated in the world of evolutionary biology."[13] Gould is wrong again. Both the "hopeful monster" theory and Gould's punctuated equilibrium theory--that are very similar which is the reason Gould came to Goldschmidt's defense--are figments of overwrought imaginations usually associated with TV script writers rather than with scientists.

Some evolutionists have alleged that Gould does not support Goldschmidt's position, but they are dishonest or uninformed. Gould wrote, "I wish to defend Goldschmidt's postulate that macroevolution is not simply microevolution, extrapolated, and that major structural transitions can occur rapidly without a smooth series of intermediate stages."[14] **That** should settle that argument.

Aren't you ladies thrilled to know that these silly scientists are wrong about a bird being born to a reptile? After all, you can imagine what might happen the next time you get pregnant! **You** could be the beginning of yet another species, even higher than man or another "hopeful monster." But rest assured it will not happen, nor did it happen millions of years ago.

## NONSENSE

If I believed such a silly theory I would never admit

it, and would surely never promote it to others. Those evolutionists are writing and speaking around the country trying to convince sane people that a reptile's egg hatched a bird, without making themselves look like **turkeys**--without success. But those "turkeys" are brainwashing little kids as well.

The silly theory is being taught to our childen and was endorsed by the American Association for the Advancement of Science in **The Wonderful Egg** (Ipcar, 1958). It was also recommended by the American Council on Education and the Asssociation for Childhood Education International. The book begins with a mother dinosaur laying an egg and asking, "Did a mother dinosaur lay that egg to hatch into a baby dinosaur?" The book answers "no" to various kinds of dinosaurs. Then comes the climax: "It was a wonderful new kind of egg." And what did the dinosaur egg hatch into? "It hatched into a baby bird, the first bird in the whole world. And the baby bird grew up...the first beautiful bird that ever sang a song high in the tree tops...of long, long, ago."[15] That's called, "brainwashing."

The Apostle Paul wrote of these foolish people in II Thessalonians 2:11-12: "...God shall send them a strong delusion, that they should believe a lie: That they all might be damned who believe not the truth...."

So the desperate promoters of the outrageous hopeful monster theory (or "punctuated equilibrium" to impress the gullible) believe such nonsense because there are **no transitional fossils** linking one species to another. Their embarrassing and tenuous theory seems to "get them off the hook," but we will not let them off the hook because genetics, paleontology and especially geology all disprove evolution and support creation, as we will see in the following chapters dealing with the universal Flood.

## CHAPTER ELEVEN

# WAS NOAH'S FLOOD WORLDWIDE?

According to the Bible, God created the universe, the world, and man in six literal days less than 10,000 years ago! Adam and Eve were real people who sinned against God and suffered the results of sin--judgment as we do today. They were forced from the Garden of Eden and had to work for their daily sustenance. It was man's darkest hour as he was chased from Eden to begin his long, lonely, and licentious trek across the centuries. During that trek he has gone **down** not up! Man did not start on the bottom and work his way to the top, but at the top and worked his way to the bottom!

Before the Fall, there was a perfect environment: there were no extremes in temperature, no storms, no disease, no sorrow, no death. After Adam and Eve's disobedience, they had to fight the bugs, beetles, and beasts to make a living. Sin sure had its consequences! It still does.

As Adam and Eve had children and grandchildren, some did not walk with God and became sinful to the point that God decided He would destroy the world and everything therein except those who obeyed Him. He decided to accomplish this result with a worldwide flood.

## SUPPORTERS OF UNIVERSAL FLOOD

Critics of the universal Flood (always wearing the

garb of scholars, **never** the rags of prejudice), refuse to accept the Flood as being universal although Christ taught it in Matthew 24:37-39: "But as the days of Noe were, so shall also the coming of the Son of man be. For as in the days that were before the flood they were eating and drinking, marrying and giving in marriage, until the day that Noe entered into the ark. And knew not until the flood came, and took them all away; so shall also the coming of the Son of man be." I'm content to stand where Christ stood! How about you?

The Apostle Peter also taught a universal Flood in II Peter 3:6: "Whereby the world that then was, being overflowed with water, perished." But then those of us who believe the Bible realize that it doesn't matter who taught it, if its taught **any** place in the Bible, it is true. Genesis 7 clearly teaches the Flood was an actual event, and it was universal. If you destroy a part of the Bible, you destroy the whole. So Genesis 7 is all we need to believe the Flood was universal.

Byron Nelson documented that the leaders of the early church believed in a universal Flood. Philo, not a Christian, but a Jewish leader, was born about 15 B.C., and wrote, "For every part of the earth sunk beneath the water, and the entire system of the world became (what it is not lawful either to speak or to think) mutilated, and deformed by the vast amputation."[1] So a first century Jewish historian believed the flood was global.

Chrysostom, was Bishop of Constantinople, and was considered the greatest preacher of the early church. He said, "The Deluge was the common wreck of the inhabited land; the cataracts were opened, the abyss flowed out again, and everything was water: the visible things were all resolved into their elements; earth no longer appeared, for all was sea. Behold now a miracle! When the earth had been obliterated...Noah went forth, preserved from the wreck....All had perished."[2]

The local flood theory is of recent origin, probably first espoused by an Italian named Quirini in 1676.[3] The Bible clearly teaches a universal Flood, and it was taught by the early church leaders as mentioned above. Augustine, the theologian of the early church also taught a global Flood.

## NOAH OBEYED GOD

Noah was a righteous man and obeyed God's command to build an ark according to His blueprint. The ark was built by Noah and his sons (no doubt with hired labor since it was a huge job). The ark was a "floater," not one that sailed. It was about 450 feet long, 75 feet wide and as high as a four story building. It had three decks with a window around the top, and one door. The ark was equivalent to 522 standard stock cars as used by railroads or equivalent to eight freight trains with sixty-five such cars in each train![4]

"Well, all right, so it was a big boat, but it wasn't big enough to hold every variety of animals in the world," says an honest critic. Well, no one says it did! What Noah did was take two of every species of land-dwelling, air-breathing creatures. He did not take every variety of dog, cat, etc. but two cats, two horses, two dogs, etc. in the boat. Only **one** of the 200 modern varieties of dogs went along! That does make for more room, doesn't it?

Furthermore, Noah probably took with him very

young animals, after all, there was no reason to take a full grown elephant, bear or ape.The smaller animals would be easier to handle, would take less space, would require less food and water, etc. The average animal in the ark was the size of a sheep, so there was plenty of room in the ark for all the animals with room to spare for living quarters, storage, etc. The size of the ark is not a legitimate criticism by Flood critics.

It is interesting to note that as late as 1858, the largest vessel of her type in the world was the P & O liner **Himalaya**, which was 240 feet by 35 feet.[5]  The ark was much larger and had proper proportions for seaworthy vessels as well. How did Noah know **that** thousands of years before the experts?

Flood critics have objected to the Genesis story because the Flood is a perfect framework for what appears in the geological record. It is a much more reasonable explanation for the present condition of the earth than the fairy tale of evolution; therefore, the critics must protect their untenable positions by seeking to discredit the universal Flood.

A world Flood account, where only a few people survived, usually because of obedience to some "god," is found in more than 200 different cultures. Of course, those stories were corrupted versions of the original Flood of Noah. The flood stories of other nations are outrageous extremes while the Genesis account is sane, logical and believable.

Let's suppose that there were not any other Flood stories from the other cultures of the world: the evolutionists would use **that** against us saying, "Surely you know enough about human nature to realize that such an event in history would be told down through the centuries, so obviously the Flood never happened."

My critics have pointed out the many flood stories from other cultures, and they make much of those nonsensical, outrageous tales, suggesting that the Bible version is just a myth like the others and derived from them. Of course we believe that the Bible version is the true one and all the others are corruptions of that actual event.

It is not at all unreasonable to expect a people to pass down something like a universal Flood that destroyed everything and everyone except a few people. It fact, it would be incredible if a culture did not have such a story, and it is expected that many of those stories would be corrupted over the years. God protected the true account by recording it in the Scripture.

The Flood would also account for the fact that recorded history goes back to, well, the time of the universal Flood, about 5,000 years ago! Well, I suppose that could be coincidental, but why do the ancient civilizations appear at the same time and very suddenly? Another coincidence?

Noah, his wife, three sons, and their wives lived in the Fertile Crescent of the Mesopotamian Valley, located between the Tigris and Euphrates rivers, an area known today as Iraq. Noah and his family were the only humans to survive the raging waters of judgment.

Bible critics tell us that the Genesis story is a camp fire story, old wives' tale, etc., that was accepted as part of Hebrew sacred literature. Of course, if that accusation is true, then John 3:16 can not be trusted either! The problem with the critics is in their hearts as well as their heads.

We are told that Noah could never have chased down all the animals and herded them into the ark, but of course, the Bible does not say that he did. It says in Genesis 7:9, "There went in two and two unto Noah into the ark, the male and the female, as God had commanded Noah." Genesis 6:20 promised that "two of every sort shall come unto thee." So, the idea that Noah spent years searching for all the animals in the world is not supported by Scripture.

Why is it so unusual for the animals to make their way to the ark at the proper time? After all, the salmon swim upstream to spawn even if they have never been there; birds fly south for the winter and back again in the spring; and there are numerous other incredible automatic responses from God's creatures, none of which can be explained by the experts!

Some scientists now believe there may be a connection between earthquakes, volcanoes, etc. and animal activity. Snakes come out of their holes, bears come out of their caves and dogs bark incessantly just before catastrophic events.

Mount Pelee on the Caribbean island of Martinique exploded in 1902 killing forty thousand people. The day before the mountain top exploded, an old black sheepherder observed his sheep in a very disturbed condition. He made his way down the mountain to the mayor's office and warned the mayor of St. Pierre that something terrible was going to happen. The mayor refused to heed the excited muttering of the old man, and the next day the island was showered with deadly, hot cinders and gases.[6] There are many similar unusual examples of animal reactions to natural, catastrophic events.

We are told that the Flood was not universal, that it was only a local flood along the Tigris River, because unbelieving scientists must never permit anything supernatural since then they would have to give an account to a personal God whom they have rejected and refused for a lifetime. However, you don't have to be too bright to realize that if it were only a local flood, there would be no need to have an ark! People could have simply climbed the nearest hill until the waters receded! Furthermore, it was God's purpose to destroy a degenerate race, and His purpose could not have been realized by a local flood that carried off only a few thousand people. No, the local flood theory doesn't **hold water**!

God promised three times after the Flood that He would never wipe out "everything living" and "all flesh" again with such a Flood, but if the Flood was local, then God did not keep His promise because millions of people have died in local floods since Noah's day. For those who believe the Bible, there is no possibility that the Genesis Flood was a local flood.

Another reason for a universal Flood was to kill all land-dwelling, air-breathing animals on the earth, (Genesis 6:17) and since they had scattered all over the world, the world-wide flood was essential to accomplish God's desire. If the Flood were only local, there would be no purpose to take animals on the ark.

Christ's warning to future generations about coming judgment on the basis of what happened to Noah's generation would be pointless if a portion of the human race escaped the Flood.[7]

This fact is made clear in Genesis 6:13 which says,

"The end of all flesh is come before me; for the earth is filled with violence through them; and, behold, I will destroy them with the earth." So, it is clear, if you believe the Bible, that the Earth was destroyed by water as geology clearly shows. There is no problem with geology, it is **theology** that is the problem. Geology supports a universal Flood, and it is, as Whitcomb and Morris said, a "mighty evangelistic witness to a world in rebellion against its Creator."[8]

The Apostle Peter taught the universal Flood in I Peter 3:20: "Which sometime were disobedient, when once the long-suffering of God waited in the days of Noah, while the ark was a preparing, wherein few, that is, eight souls were saved by water."

Again, in II Peter 2:5, he wrote, "And spared not the old world, but saved Noah the eighth person, a preacher of righteousness, bringing in the flood upon the world of the ungodly." So the Bible is clear: the Flood was worldwide, and killed every land-dwelling, air-breathing creature not in the ark.

## RESULTS OF WORLD FLOOD

The Genesis Flood produced driving rain, surging oceans, raging streams, tidal waves, seaquakes, earthquakes and other violent reactions. The earth heaved, basins sank, mountains were raised and the waters dropped fossil-laden sediment over most of the earth. Results of all the above can be seen everywhere.

The major rivers of the world have all been at much higher levels. In Washington, D.C., the waters of the Potomac Basin at one time flowed up to 265 feet **above** present levels. The same is true of the Spokane River in Washington state and of other rivers across the Earth.[9]

All animals and billions of fish were destroyed in the Flood and were covered with sediment thereby producing the fossils that evolutionists talk so much about. However, they try to convince sane people that the animals and fish were covered slowly over billions of years, but the fossil record does not support that theory.

Even evolutionary geologist, Carl Dunbar, admitted that almost all fossils were inundated by catastrophic events such as floods. He described how a creature dies, and its body is torn apart by scavengers. If it escapes, it will decay. He adds that if the carcass is buried under sediment, it is preserved and safe from predators.

Dunbar wrote, "Water-borne sediments are so much more widely distributed than all other kinds, that they include the great majority of all fossils. Flooded streams drown and bury their victims in the shifting channel sands or in the muds of the valley floor."[10]

How does a dead fish lie on the bottom of a lake for hundreds of years until it is fossilized? Of course, that doesn't happen in real life, only in the fertile and desperate minds of evolutionists. The animals, plants and fish were fossilized almost immediately as the flood waters receded. The surging rivers carried sediment along with animals, trees, etc. and deposited them in layers (strata).

Often large, fighting, writhing animals were dropped on a stratum, and then a heavy layer covered them with enormous weight, crushing the animals. The universal Flood resulted in millions of fossilized animal remains found in massive graveyards with their bones encased in rock. Their soft body parts quickly decayed leaving either their bones or impressions of where the bones had been as silent testimonials to sudden destruction. Fossilized fish have been found numbering in the thousands of millions in an apparent state of panic without any evidence of predator attack.

In Agate Springs, Nebraska, there is a stratum where thousands of bones from fossilized animals are jumbled together, indicating unusual transportation and rapid burial.

Dr. Duane Gish reports on the burial conditions of creatures in the "Cleveland shale" in Ohio. All kinds of fish, mostly sharks, were found from five feet to fifty feet below the surface. Five foot sharks were pressed flat from top to bottom to the thickness of a quarter of an inch! Their entombment clearly shows that they were trying to swim in sediment-laden water, but were unable to swim and were finally crushed by the settling sediment.[11]

Geologist Hugh Miller, described the Devonian sedimentary deposit covering much of England: "At this period in our history, some terrible catastrophe involved sudden destruction of the fish of an area at least a hundred miles from boundary to boundary...[the area has been] strewed thick with remains, which exhibit unequivocally the marks of violent death. The figures are contorted, contracted, curved; the tail in many instances is bent around to the head; the spines stick out, the fins are spread to the full, as in fish that die in convulsions..."[12] **That** is the result of the universal Flood.

What happens when a fish or sea creature dies today? We have all seen the results: the body floats on the surface of the water or sinks to the bottom where it is devoured quickly by other fish. But the fossil fish are often found very well preserved in sedimentary rocks. I. Velikovsky, author of **Earth in Upheaval**, says that entire shoals of fish over large areas "...numbering billions of specimens, are found in a state of agony, but with no mark of a scavenger's attack."[13]

Dinosaurs and other creatures have been unearthed in positions that suggest violent and sudden death. Many entire skeletons of duck-billed dinosaurs have been excavated with the head thrown back as if in death throes. Various kinds of creatures seemed to instinctively know that death was imminent.

Evolutionists have had to scramble madly to explain why so many land animals died violently and suddenly in water, but the answer is simple: they were buried quickly in the sediment of the waters during the universal Flood.

Remember that all creationists recognize the billions of fossils that have been uncovered; however, you must understand that fossils do not come stamped with their birthday! Three questions arise: how old are the fossils, how did they die, and how were they covered?

There are remnants of a worldwide Flood all over the Earth. Marine crustaceans have been found in the Alps, twelve thousand feet above sea level; hippopotamuses have been unearthed in England, and about twenty mammoths were dug up near the Neckar River in Germany. Hordes of dinosaurs, mixed with other creatures that did not share the same environment have been found buried together.

Huge fossil beds have been discovered where **billions** of creatures' remains were buried together. One such place is the Karoo Formation of South Africa that contains the remains of 800 **billion** vertebrates--such as reptiles.[14] And many of those creatures did not share the same environment, yet they were buried together! Evidently they did have one thing in common: Escape the flood.

There are mass burial sites throughout the world that are packed with fossils from various widely separated current climatic zones. The Norfolk forest-beds in England contain fossils of northern cold-climate animals, tropical warm-climate animals, and temperate zone plants, and the herring fossil bed in the Miocene shales of California contain the remains of a billion fish that died within a four-square mile area. The Cumberland Bone Cave of Maryland contains remains of dozens of species of mammals from bats to mastodons, as well as reptiles and birds from the Arctic region to tropical zones. Evolutionary teaching would require that these creatures were carried great distances from their place of origin to their final burial locations in a violent cataclysm, the universal Flood.

Alexander Catcott wrote of "incontestable proofs of the Deluge over the face of the earth." He reported that it seemed to have "...happened last year...Search the earth; you will find the moose-deer, native of America, buried in Ireland; elephants, natives of Asia and Africa, buried in the midst of England; crocodiles, natives of the Nile, in the heart of Germany; shell-fish, never known in the American seas, together with the entire skeletons of whales, in the most inland regions of England..."[15] How did such diverse creatures get buried thousands of miles from their normal environments except by a devastating universal Flood?

Another scientist, geologist Benjamin Silliman, who was head of the Geology Department at Yale University, believed the Flood was world-wide. He said, "Geology fully confirms the Scriptural history of the event...Whales, sharks, and other fishes, crocodiles and amphibians, the mammoth and the extinct elephant, the rhinoceros, the hippopotamus, hyenas, tigers, deer, horses,...are found buried...under circumstances indicating that they were buried by the same catastrophe which destroyed them: namely a sudden and violent deluge."[16]

Silliman also tells of a whale's skeleton on top of the mountain Sanhorn which is 3,000 feet high! He said nothing could have conveyed the whale to that height except a deluge.

How did elephants and other large creatures get to the highest mountain tops unless carried there by the Flood? Can you imagine an elephant climbing a mountain even if it wanted to?

Some fossil burial grounds are high on mountain tops such as those in Sicily on Mt. Etna where two caves are packed with the bones of thousands of giant hippopotamuses! On the island of Malta, there are lions, tigers, birds, beavers, hippopotamuses, and foxes all jumbled together.[17] How did so many different kinds of animals die and get buried together except by a massive flood that extended past the tops of the mountains?

The Flood of Genesis dropped a layer of sedimentary deposits over three-fourths of the earth, and in the Tibetan Plateau, there are 750,000 square miles of it thousands of feet thick--at an elevation of three miles![18]

Leaves have been found in lignite beds in Germany that were deposited and preserved in a "fully fresh condition." The chlorophyll, usually easily destroyed, is so well preserved that it is possible to recognize the alpha and beta types. Heribert Nilsson says that the "...incrustation must therefore have been very rapid."[19]

What could have thrown an amalgamation of fish, animals, and plants together except a tumultuous world Flood?

The evolutionist tells us that the thousands of feet of strata were laid down over billions of years, one grain at a time, but all the evidence tells us that it happened very quickly. A good example is the upside down whale! And by upside down, I mean from head to tail!

In a quarry in Lompoc, California, in 1976, workers discovered the fossil skeleton of a baleen whale estimated to be eighty feet long. It was fossilized in an upside down position in billions of diatom shells. The fossil stands on its tail in the quarry, and is being exposed as the diatomite is mined.[20]

Now, why didn't the whale rot soon after its death or why didn't predators devour it? (And how did it get on its tale?) Evolutionists tell us the fossils develop over long ages, but the whale would have decomposed in **days** and its skeleton in months. The whale was buried quickly during the flood without any chance of being devoured by predators. (In fact, the predators were trying to escape the destruction of the Flood.)

It has been said that if evolution is true, the fossils would reveal it; if it is fraudulent, they will refute it. It is obvious to all thinking people that the fossil record does not support evolution but demands a world Flood.

How could millions of land and sea creatures from various parts of the world, be buried together if not as a result of a world Flood? The strata with fossils, large and small, demand a churning, violent flood.

Obviously, the numerous strata around the world are the result of a slowly settling Flood. Byron Nelson wrote, "A stream moving from side to side, or alternately fast and slow over one area for several hours or days would inevitably sort its material and deposit them in horizontal layers one above another, and we do not see what else could possibly do so."[21]

Noah's Flood answers the major problems concerning the geologic condition of the earth, but unbelieving scientists' knees begin to jerk incessantly when a universal Flood is discussed. They would say the universal Flood is too preposterous to be believed, even by the ignorant folk of Noah's day. But, of course, **they** missed the boat! Just like evolutionists today.

# THE GEOLOGIC COLUMN
## (According to Evolutionists)

| ERAS | PERIODS | BEGAN MILLIONS OF YEARS AGO | EVOLVING LIFE |
|------|---------|------|------|
| **CENOZOIC** | *PLEISTOCENE* | *1* | *MAN* |
| | *PLIOCENE* *MIOCENE* *OLIGOCENE* *EOCENE* | *60* | *HOMINIDS* *APES* *MONKEYS* |
| **MESOZOIC** | *CRETACEOUS* | *130* | *PRIMATES* *WHALES* *MARSUPIALS* *MAMMALS* |
| | *JURASSIC* | *180* | *FLOWERING PLANTS* *BIRDS* |
| | *TRIASSIC* | *220* | *DINOSAURS* *CONIFERS* |
| **PALAEOZOIC** | *PERMEAN* | *280* | *REPTILES* |
| | *PENNSYLVANIAN* (*CARBONIFEROUS*) | *310* | *PRIMITIVE REPTILES* *SWAMP FORESTS (COAL)* |
| | *MISSISSIPPIAN* (*CARBONIFEROUS*) | *350* | *AMPHIBIANS* |
| | *DEVONIAN* | *400* | *SHARKS* *BONY FISH* *FERNS* |
| | *SILURIAN* | *450* | *INSECTS* *MOSSES* |
| | *ORDOVICIAN* | *500* | *JAWLESS FISH* *JELLY FISH* |
| | *CAMBRIAN* | *600* | *MOLLUSCS* *SPONGES* *INVERTEBRATES* |
| **PROTEROZOIC** **ARCHAEOZOIC** | *PRECAMBRIAN* | *3000* | *BACTERIA* *ALGAE* |

# CHAPTER TWELVE

# MORE RESULTS FROM THE FLOOD

One of the most incredible results of Noah's Flood was the entombment in ice (and sedimentary strata) of about five million mammoths along northern Siberia and Alaska. Some of the islands in the Arctic Ocean off the coast of Siberia are littered with bones of animals! One can walk along and see many elephant tusks protruding from the ground. Sheep, rhinoceroses, horses, oxen and bison remains seem to be everywhere.

On mainland Siberia the whole coastline is embedded with elephant remains, and wherever rivers have cut their beds, the ground is said to "...consist entirely of mammoth bones."[1]

The **New York World** reported an exciting discovery on June 1, 1930: "Fairbanks, Alaska, May 31--Well preserved skeletons of lions, elephants, mastodons, horse, bear, bison, moose and various rodents have been uncovered here by the two huge gold-dredging machines cutting into the fossil-filled muck which overlays the gold bearing gravel. For more than a hundred years engineers, prospectors and miners in Alaska have encountered tusks, bones and skulls of extinct Pleistocene animals, exposed in the banks of rivers...The gold mining dredges now being pushed deep into the old earth have years of work ahead, and already the scientific results greatly surpass expectation..."[2]

Those were skeletal remains, but in northern Siberia and Alaska, the **flesh** is sometimes preserved! Think of it! In that cold, treeless, inhospitable land of Northern Siberia, locked in by snow and ice, where only a few polar animals live, there are perfectly preserved remains of thousands of elephants and other beasts! (Further south, usually the bones are found rather than preserved flesh.)

Some of that flesh has been eaten by wolves, dogs and bears. Many of those mammoths, (extinct for millions of years, according to evolutionists) were killed and frozen so quickly they have been found with grasses, bluebells, buttercups, sedges and wild beans in their mouths!

One such report described a mammoth found near the Beresovka River in Siberia. It was "...peacefully grazing on summer buttercups in late July and within one half hour of ingestion of his last lunch, he was overcome by temperatures in excess of -150° F...Whatever climatic upheaval caught him, permanently changed the climatic conditions of the tundra."[3]

Ivan T. Sanderson, a well-known zoologist, took the position that these incidents were not Flood related until he became aware of the above incident. Sanderson studied the "Beresovka mammoth" after it was taken to the Leningrad Museum, and almost completely reversed his position!

Sanderson wrote, "First, the mammoth was upright, but it had a broken hip. Second, its exterior was whole and perfect, with none of its two-foot long shaggy fur rubbed or torn off. Third, it was fresh; its parts, although they started to rot when the heat of fire got at them, were just as they had been in life; the stomach contents had not begun to decompose. Finally, there were buttercups on its tongue."[4]

How could hundreds of thousands of these huge wild beasts be frozen so quickly that they couldn't swallow their food or lie down? A bone-freezing wind must have fallen on them like a noxious fog over thousands of square miles throughout the north polar region. Alaskan sled dogs have been out in blizzard conditions for days in temperature 80 degrees below zero without freezing, yet those huge furry animals were frozen to death before they could swallow the food in their mouths! Was God leaving a frozen memorial to the universal Flood to be discovered in our time?[5]

It is a fact that the Siberian soil is full of ancient beasts, some fossilized bones and others quick frozen but in good physical condition (other than being dead). All this indicates that Siberia was at one time a land of lush green vegetation. In fact, the vegetation would have been fantastic to feed the estimated five million mammoths (plus all the other animals) that roamed Siberia and down into Alaska.

So Siberia, at one time, had a warm, lush climate that permitted flowers and other plants to grow. After all, if some of the frozen mammoths died with warm weather plants in their digestive system (and in a few cases, in their mouths) then Siberia **was** tropical at one time. The same holds true for dinosaurs. Since they were coldblooded creatures, they required warm climate, and their remains have been found on every continent! The tropical weather was, at one time, worldwide.

The Arctic explorer, Baron Toll, found the remains of a saber-toothed tiger over 600 miles north of the Arctic Circle in the New Siberian Islands! At the site he also found a preserved 90-foot plum tree with green leaves and ripe fruit on its branches! That is in an area where the only vegetation that grows is a one-inch high willow![6]

Dr. Jack A. Wolfe supported the northern warm weather theory in a U.S. Geological Survey Report as reported by John Whitcomb. He told that "...Alaska once teemed with tropical plants. He found evidence of mangroves, palm trees, Burmese lacquer trees, and groups of trees that now produce nutmeg and Macassar oil."[7]

So it's a fact that Siberia was tropical at one time with delightful fruits, vegetables, and grasses that supported a huge assortment of small and large creatures. And those creatures all died at the same time, killed by the same event and were buried the same way. Astounding!

## BIBLE CRITIC SPEAKS

Let's hear what a **Bible critic** says about all this. Henry Howorth is a geologist who said the Bible had "no authority whatever." When you realize he is a Bible **critic**, his statement about these frozen fossils is astounding! "When nature puts a term to an animal's life in her normal way, it is exceedingly seldom she does so when the animal is young. Animals do not die naturally in crowds when young, and yet we find remains of quite young animals abounding in all classes from mammoths to mice. How are we to account for this fact, save by summoning an abnormal cause? How again can we account for the fact that the mummied animals found in Siberia seem to have been in robust health, stout and strong? Is this, again, consistent with a natural death? Again if the death was natural and other carnivorous animals abounded, would the corpses be left to the useless duties of decay, as they must have been since the bones are ungnawed, and (when the flesh is preserved) the flesh uneaten?"[8]

He continues his amazing endorsement of flood-geology: "One cause, no doubt, of the scarcity of remains of animals which are dying at present, where animal life abounds, is the diligence of the scavengers. What were they doing...to pass by these myriads of corpses, and in so many cases not to leave a toothmark anywhere, and in fact to leave their own bones with the rest? Surely this points clearly and unmistakably to the fact that the animals, or the greater part of them, died together. If the remains were the silent chronicles of centuries of time and generations of life (as uniformitarians suppose) we should assuredly have found some or a large portion of the bones would have been gnawed, but this is not the case, and it points strongly to their death having been more or less simultaneous."[9]

Howorth goes on to deal with the possibility of a pestilence killing the various creatures, but then asks what disease would sweep away all forms of animal life? He also suggests that a pestilence would not sweep over two whole continents, nor would it, after killing the creatures, bury them in the same place!

He then points out that animals of different species don't come together to die of natural causes, nor does the lion come to take his last sleep with the lamb. The animals were all in the same state of preservation indicating they all experienced the same demise. It is also true that they were all buried at once, leaving no time for decay or weathering.

Now Howorth gets to the heart of the problem: What killed the various creatures? He builds his case for death by water: "We want a cause that should kill the animals, and yet not break to pieces their bodies, or even mutilate them; a cause which would in some cases disintegrate the skeletons without weathering the bones. We want a cause that would not merely do this as a widespread plague or murrain might, but one which would bury the bodies as well as kill the animals, which would take up gravel and clay and lay them down again, and which could sweep together animals of different sizes and species, and mix them with trees and other debris of vegetation. What cause competent to do this is known to us? Water would drown the animals and yet would not mutilate their bodies. It would kill them all with complete impartiality, irrespective of their strength, age or size. It would take up clay and earth, and cover the bodies with it...Not only could it do this, but it is **the only cause known to me capable of doing the work on a scale commensurate with the effects we see in Siberia.**"[10] (Emphasis added.) That is excellent support for a universal Flood by an unbeliever!

## THE CAUSE

The evolutionists are silent about the fossil problem in Siberia and Alaska, although Darwin was aware of it, and confessed that he didn't have the answer.[11] Lyell suggested that they were all caught swimming during a sudden cold snap, but only a world flood offers a reasonable explanation for those unusual fossils.

It is known that people have been aware of the frozen mammoths and skeletal remains for at least 2,000 years. Since 900 A. D. men have collected the ivory tusks and sold them to China, Arabia and Europe. Generations of men have made a good living selling the tusks of the extinct monsters.

Thoughtful people have asked how the mammoths and other animals were not only killed but "quick frozen" so effectively since there is nothing like that happening today. It is believed by most creationists that the earth, before the Flood, was consistently warm with no violent storms, volcanos, earthquakes, etc. High winds and violent storms would not be possible since they are produced by variation of temperatures, and the temperatures would have been rather constant over the Earth. In fact, Genesis 2:5-6 teaches that there was no rain before the Flood. A mist came up from the Earth and watered it nightly.

Creationists believe that the ocean covered the whole earth for a time, but there was another layer of water **above** the earth in the form of water vapor. This is called, the "canopy theory." It is noteworthy that ancient Hebrews viewed the sky as a thin dome holding back the water "above" (rain) from the water "below" (the sea). This canopy above kept the Earth from the violent storms that we now experience, and it assured a universal and consistent tropical temperature. The canopy would have permitted heat from the sun to get through to the Earth without harmful rays, thus an explanation for the long lives of early mankind as reported in the Bible. It might also account for the much larger animals that roamed the Earth.

The hornless rhinoceros grew to be nearly thirty feet long and seventeen feet high, and cockroaches were 12 inches long! Dragonflies today have about four inch wing spans, but in ancient times they were up to three feet across![12] The remains of giant bears have been found that make the Kodiak bears look like their growth was stunted!

Birds that stood ten feet high with wing spans of 25 to 30 feet have been unearthed. (They laid gallon-sized eggs almost eleven inches in diameter!) Turtles have been found that were 12 feet long![13] Of course we are all aware that some dinosaurs weighed forty tons! The perfect environment, plenty of food, and lack of dangerous sun rays could have contributed to these huge specimens.

It is believed that God collapsed the canopy above the Earth, and great quantities of water fell as the floodgates of heaven were opened. It has been argued that the Earth could not have contained the additional water (that eventually covered the highest mountains); however, Psalm 104:6-9 and Genesis 8:3 suggest that the ocean basins were deepened to make room for the additional water that had been the canopy. Geologists have established the fact that present ocean floors are several thousand feet lower than in the past.

Critics then attack from the opposite direction and tell us that there was not enough water available, even with the canopy, to cover the whole earth, but they assume that the present topography was here during the Flood. That assumption would require a flood to cover the highest mountains, but some of the high mountains were the **results** of the Flood so less water was required than critics say.

There was water from the heavens, water from inside the earth and water from the oceans, and we now know that the oceans are deeper than Mount Everest is **high**! It has been estimated that if the ice caps on Greenland and Antarctica were melted, the oceans would be raised 200 feet above their present level.[14] So there was plenty of water to accomplish God's plan to inundate the world with a universal

Flood.

Psalm 104:5-9 tells us that is **exactly** what happened: "Who laid the foundations of earth, that it should not be removed for ever. Thou coveredst it with the deep as with a garment: the waters stood above the mountains. At thy rebuke they fled; at the voice of thy thunder they hasted away. They go up by the mountains; they go down by the valleys unto the place which thou hast founded for them." That's a terrific description of the Flood.

Richard Carrington explained the raising of major mountains in his book, **The Story of Our Earth**: "In the Old World, the Alpine and Himalayan mountain systems were upraised from the bed of the sea by the compression of the land masses on either side. It is awe-inspiring to contemplate the physical forces that caused these stupendous bodies of matter to be slowly lifted, and then ride over the neighboring shoreland in huge waves of twisted rock."[15] Of course, it didn't happen slowly, but swiftly during the Flood of Noah.

When "all the springs of the great deep burst forth," indicating the Earth's crust had broken apart by earthquakes and volcanic activity, the Earth was flooded and violent weather patterns began. With the canopy destroyed, the world experienced extreme weather with the sun and winds developing violent Earth-changing weather systems all over the Earth. Huge amounts of water fell onto the Earth and gushed from the Earth to fill the land and the seas. This would explain the violent deaths of millions of animals in swirling, shifting, churning flood waters. With the protective canopy gone, the Earth then reacted to the effect of the sun and winds resulting in violent changes of temperature. Huge quantities of volcanic dust would have produced a precipitous drop in temperature and at this moment the mammoths and dinosaurs were destroyed and many were quick frozen.

## ROCKS PROVE WORLD FLOOD

The world was covered with water, earthquakes rumbled from pole to pole, seaquakes disturbed the oceans and seas, violent storms whipped across the face of the globe and massive waves churned up violent whirlpools as lightning streaked across the foreboding heavens. (Noah and his family were safe inside the ark, built according to God's plan.) During the year of flood, the churning water laid down sediment over three-fourths of the Earth. (Only one-fourth of the rock is volcanic rock.)

Proof that the Flood covered our highest mountains is seen in the sedimentary rock that covers those mountains. Sea fossils are found on all the mountains of the world! Furthermore, geologists have found a field of pillow lava on Mt. Ararat at the 14,000-15,000 foot level. Pillow lava is formed **only** under water. Got any answers Steve? How about you Niles? Call me collect for your answers.

Evolutionists tell us that the sedimentary strata were not laid in a short period of time, but over long ages lasting billions of years! They would have us believe that the strata were formed when, billions of years ago, the land was eroded by rivers, rain and wind. The dirt from the erosion was then washed down to the seas by the rivers and spread out on the bottom of the oceans. Later, the ocean floor raised and the water receded to other parts of the Earth revealing the ocean bottom (now exposed land areas). This has been happening, according to the story teller, for billions of years. Of course, there is no proof for that theory. An honest look at the facts of geology supports a world Flood, but that can't be accepted by evolutionists.

The evolutionists are faced with millions of fish that are buried in the rocks of the American Rocky Mountains, England, Germany, Switzerland, etc. and also the way elephants and rhinoceroses  and ancient mammoths are buried in Siberia, Alaska and other parts of the Earth; yet they still refuse to admit a universal Flood. They weakly tell us that there were many local floods that probably produced the unusual results! Sure!

The great coal seams of the world were created during the Flood from the lush forests that covered the tropical Earth, not as evolutionists say, slowly formed over billions of years. It has been commonly believed, until recently, that coal seams are formed as a forest dies, are covered by sediment, then another forest grows on top of that same spot, then dies in a few million years, to be continued over and over. We are not told how the forests grow in the very same spot many times (as many seams of coal are found).

There are places where fifty different forests would have had to grow to produce fifty different seams of coal! Strange that, over millions of years, the forests kept growing in the **same** spot, producing fifty clearly defined coal seams one on top of the other and not get out of line with each other![16]

There is another problem the evolutionists can't answer. If the coal seams were developed over millions of years,  they would have been exposed to the elements all that time, yet there is never any evidence of erosion! The seams of coal are smooth and even, never showing effects of being exposed to ages of weather!

It is obvious to the unbiased that massive amounts of vegetation were carried by swirling waters and dumped in a location. Then, a layer of dirt and mud washed over that layer of vegetation and was deposited followed an hour later by another layer of vegetation. All this took place in a very short period of time. The vegetation was pressed by the various layers of dirt and then more vegetation, producing the coal seams in the eastern part of America and other parts of the Earth.

Evolutionists deny the Flood (since it proves the Bible, and the Bible reveals God), but they can't refute the evidence of the Flood. Creationists and evolutionists don't disagree on geology, but the **interpretation** of geology.

The critics of Flood geology have a massive problem with the fossils found in the rock and coal strata. Often, upright trees are found through different seams of coal (that took millions of years to form according to evolutionists). A good example of this is found in the Craigllieth Quarry in England where an eighty foot tree was discovered that intersected up to twelve different strata of limestone! How could a tree live millions of years so that the strata could be laid around it? Why didn't it rot as trees do today? Evolutionists are strangely silent.

Large boulders have been discovered in coal indicating that they were transported there by water--during the Flood. When else could it have happened?

Coal could not have required millions of years to produce because of the man-made articles found in the seams! If coal seams were formed millions of years before man, evolutionists must explain how recent objects got into the coal! (The same is true of the sedimentary strata.)

Mrs. S. W. Culp broke open a chunk of coal and found in it a ten inch eight-carat gold chain, so the coal had to be formed after a human had made the chain![17] The jawbone of a child has been found in coal that is supposed to be thirty million years old![18] However, that can't be possible if you believe the evolutionist fairy tale. Evolution is a farce, and people who believe it are fools.

It has been commonly believed that long ages are required to produce coal, but new information shows that coal similar to anthracite can be produced rapidly by a short heating process. Dr. George R. Hill of the College of Mines and Mineral Industries of the University of Utah has demonstrated that formation of coal from woody material can be done in a very short period of time.

Hill seemed to support rapid formation of coal through pressure and heat when he said, "These observations suggest that in their formation, high rank coals, i.e., anthracite and low volatile bituminous,...were probably subjected to high temperature at some stage in their history. A possible mechanism for formation of these high rank coals could have been a short time, rapid heating event."[19]

Dr. Paul Ackerman, of Wichita State University, confirmed the above when he wrote that "scientists now know that under proper conditions coal and oil can be formed very quickly. Coal has been synthesized in the laboratory in about twenty minutes and oil in about two hours."[20] Well, I would call that "quick," unlike evolutionists who insist on millions of years to develop coal and oil.

When one looks at the hills, streams, rocks and other formations of the earth, he thinks of a massive catastrophe that had to take place. It did. It is called Noah's Flood, and it explains many of the questions that evolutionists have not been able to answer for generations. The universal Flood of Genesis has been discussed, debated, denigrated, and denied but never disproved.

# CHAPTER THIRTEEN

# WHERE IS NOAH'S ARK?

The Genesis Flood was the greatest, most impacting event in the history of the world other than the death and resurrection of Christ; however, many scientists refuse to consider a universal Flood as the answer to present geological conditions. They refuse to consider the Flood because it would suggest that the Bible is true (gasp!) and judgment is ahead!

Likewise, most scientists refuse to admit that Noah's ark might still exist somewhere on an isolated mountainside in Turkey. In fact, that possibility makes many of them shudder. Dr. Melville B. Grosvenor, the late editor of the **National Geographic** magazine, aptly observed, "If the Ark of Noah is ever discovered, it would be the greatest archaeological find in human history, the greatest event since the resurrection of Christ, and it would alter all the currents of scientific thought."[1]

Can you imagine the impact on the world if tomorrow's newspapers announced the discovery of Noah's ark along with photos? Television crews would drop on the mountain like a chicken on a June bug to provide eyewitness accounts. They would show the ark's animal cages, living quarters, eating utensils, etc. The unbelieving scientists would soon gnash their teeth in agony.

What a shock to the pompous evolutionists to know that the Genesis account **is** reliable after all! Well, that day will come in my opinion since the evidence indicates that Noah's ark is perched on a mountainside in Turkey waiting to be rediscovered. I say, "rediscovered" because many people have seen and touched the ark!

## THE ARK ON ARARAT

The ark appears to be located at about the 14,000 foot level on Mount Ararat buried beneath many feet of ice and snow. This boat has been mentioned as being on that mountain by explorers and historians as early as 700 B.C. Pilgrims climbed Mount Ararat 700 years B.C., and scraped tar off the sacred vessel from which they fashioned good luck omens.

In about 30 B.C., Nicolaus of Damascus, the biographer of Herod the Great, wrote a vast universal history from earliest times to the death of Herod, and in his 96th book, he dealt with the ark. He wrote, "There is above the country of Minyas in Armenia a great mountain called Baris, where, as the story goes, many refugees found safety at the time of the flood, and one man, transported upon an ark, grounded upon the summit, and relics of the timber were for long preserved; this might well be the same man of whom Moses, the Jewish legislator wrote."[2]

Josephus was a famous Jewish historian who wrote about 100 A.D. in his **Antiquities of the Jews**, "...the Ark landed on a mountaintop in Armenia. The Armenians call that spot the landing place, for it was there that the Ark came safe to land, and they show the relics of it even today."[3]

St. Theophilus of Antioch, formerly a pagan, became the sixth bishop of Syrian Antioch in 180 A.D., and wrote of the Flood and the ark: "Moses showed that the flood lasted forty days and forty nights, torrents pouring from heaven, and from the fountains of the deep breaking up, so that the water overtopped every high hill 15 cubits. And thus the race of all the men that then were was destroyed, and those only who were protected in the Ark were saved; and these, we have already said, were eight. And of the Ark, the remains are to this day to be seen in the Arabian mountains."[4]

John Chrysostom lived from 345 to 407 A.D., and is considered the greatest preacher of the early church. In one of his sermons he said, "Let us therefore ask them (the unbelieving): Have you heard of the Flood--of that universal destruction? That was not just a threat, was it? Did it not really come to pass--was not this mighty work carried out? Do not the mountains of Armenia testify to it, where the Ark rested? And are not the remains of the Ark preserved there to this very day for our admonition?"[5]

Isidore of Seville, who died in 636 A.D., was the first professing Christian writer who wrote a book of universal knowledge, an encyclopedia, entitled, **Etymologies**. His book is still considered an authoritative work on the early middle ages, and in it he said, "...the historians testify that the Ark came to rest after the Flood. So even to this day wood remains of it are to be seen there."[6]

Marco Polo, the renowned explorer (who lived 1234-1324), wrote in his book, **The Travels**, "And you should know that in this land of Armenia, the Ark of Noah still rests on top of a certain great mountain where the snow stays so long that no one can climb it. The snow never melts--it gets thicker with each snowfall."[7]

Jans Janszoon Struys was a 17th Century Dutch adventurer who had an unusual life. He records how he was traveling in the area of Ararat in June of 1670, and was asked to go to a hermitage on the side of Mount Ararat to treat a monk with a serious hernia condition. He was successful in his treatment, and the grateful monk gave him a small wooden cross that was craved from wood taken from Noah's ark.

The monk asked him to give the cross to St. Peter's Church in Rome which he did with the following written testimony that he procured from the monk: "I have thought it unreasonable to refuse the request of Jans Janszoon (Struys) who besought me to testify in writing that he was in my cell on the holy Mt. Ararat, subsequent to his climb of some thirty-five miles.

"This man cured me of a serious hernia, and I am therefore greatly in his debt for the conscientious treatment he gave me. In return for his benevolence, I presented to him a cross made of a piece of wood from the true Ark of Noah.

"I myself entered that Ark and with my own hands cut from the wood of one of its compartments the fragment from which that cross is made.

"I informed the same Jans Janszoons in considerable detail as to the actual construction of the Ark, and also gave him a piece of stone which I personally chipped from the rock on which the Ark rests. All this I testify to be true--as true as I am in fact alive here in my sacred hermitage.

Dated the 22nd of July, 1670, on Mt. Ararat--

Domingo Alessandro of Rome"[8]

I have presented information about the Flood and Noah's ark from some of the world's most respected authorities. Surely it can't be dismissed out of hand without some serious consideration. If so, how do we accept anything from ancient historians? The writers are people of impeccable reputation. Of course, our faith does not depend upon whether the ark is still on Ararat. We know it used to be there because the Bible is very clear on that point.

The Bible tells us that the ark landed on the "mountains of Ararat" 150 days after the Flood began. Ararat is located on the border separating Turkey from the former Soviet Union. The Persians call Mount Ararat, "Kok-I-Nou," meaning "mountain of Noah." Every village seems to support the landing of the ark on the mountain that hovers over everything.

For example, the town of Nakhitchevan in 100 A.D. was called, "Apobaterion," meaning "landing place," while nearby Temainin translates to mean, "place of the eight," which seems to refer to the number of those in the ark. The town of Sharnakh means "village of Noah," and another village called Tabriz or Ta Baris means, "the ship." I could go on and on like this, but you can understand that I am not talking of a few isolated legends, but a culture that has been immersed in the story of Noah and the ark since it landed there thousands of years ago.

When the ark landed on Ararat, Noah and his family finally disembarked, and immediately offered sacrifices to God for their safe passage. From this small family, came all the people of the Earth, and anthropologists seem to confirm that fact. Anthropologists Andrews, Hrdlicka and Osborn have clearly affirmed that North and South America, Africa and the Pacific islands were peopled by folk whose ancestors originated in central Asia.

## INCREDIBLE KNOWLEDGE

Is it accidental that the earliest civilizations of the world were in the area of Ararat, known as the "cradle of civilization?" There was Egypt, Babylon (present-day Iraq), Syria, Crete, etc. These civilizations seemed to rise out of nowhere, according to historians, but we know that they were the descendents of Noah with all the knowledge and ability of their ancestors at their disposal.

With that knowledge they built amazing structures that confound modern men today. The oldest building in the world is thought to be the Great Pyramid at Giza in Egypt, the lone survivor of the seven man-made wonders of the ancient world. In fact, it is also believed to be the largest building and maybe the most **perfect** building in the world.

This pyramid stands on 13 acres and was originally 485 feet in height. It consists of about two and a half million stones, weighing from three tons to 600 tons each! It could contain **all** the cathedrals in Rome, Milan and Florence with space left for the New York Empire State Building, Westminister Abbey, St. Paul's Cathedral and the English House of Parliament. That is a **big** building, and the foundation has not settled a fraction of an inch![9] There is not a construction company in the world that would guarantee a similar job! Question: Where did ancient men get the knowledge to perform such a task? Maybe from knuckle-walking brutes only recently out of their caves?

Baalbek in Lebanon is another example of amazing

workmanship by early man. I have visited the ruins at Baalbek five or six times in recent years, and have been astounded at the work of unsophisticated, untrained men (if we believe evolutionists). The seventy foot columns were quarried in Egypt, ferried across the Mediterranean Sea, then dragged over the mountains to be erected at Baalbek. One of the foundation stones weighs 2,000 tons, and could not be moved using modern equipment! How did ancient men get the job done?

There are many other incredible structures that are inexplicable such as a South American city 13,000 feet above sea level, the hanging gardens of Babylon, the amazing cut-stone monument at Stonehenge and hundreds of pyramids in Mexico and Central America.

All these structures and others were built by the descendents of Noah as they spread throughout the world following their departure from the ark, and their dispersion at Babel. They carried with them all the accumulated wisdom of the ancient world, and put that wisdom and knowledge to work in building the ancient civilizations of the earth.

Melvin G. Kyle, an outstanding scholar, seems to support my basic premise as stated above: "The theory of this location [i.e. valley of the Euphrates] of the point of departure of the dispersion of the race, as indicated by the record of the Bible and by facts ascertained through research, is all but universally held....Wherever it is possible to trace back lines of migration of the early nations mentioned, or to gather notes of direction from the traditions of various peoples, it is always found that the ultimate direction is toward a comparatively small area in western Asia."[10] Of course, that's so surprise to Bible-believers.

All right, so the race of men that was fathered by Noah, started someplace in Asia. That seems to be without question. That still doesn't mean the ark is still on a Turkish mountainside. Right, and we Christians don't require that; but if it is there, it would sure make an impact on the evolutionists--and others.

But why, with modern equipment, don't they simply climb the mountain or approach it by helicopter and settle the issue? Well, it isn't that simple. Turkey is near the former Soviet Union, a nation that was paranoid about their security, so it was very difficult to get permits to climb the mountain from the Turks lest they antagonize their powerful neighbor.

The biggest problem is the weather. Ararat is always covered with glacial ice and snow, and the summit is almost always shrouded with heavy clouds, producing blizzard conditions throughout most of the year. The glacial ice cap is over 200 feet thick, covering 20 square miles at about the 13,500 foot level. Extending from the main glacier are twelve finger glaciers.

John Morris, a veteran Ararat climber, said, "Usually, in the morning, the mountain is crystal clear and it's a beautiful sight. But almost every day about ten o'clock, the haze sets in and by three o'clock, there's a storm on Mt. Ararat. At lower elevations, it's a severe thunderstorm while at higher levels it often results in a blizzard."[11]

Winds whip along the mountain at 150 miles per hour sending huge boulders down the slopes, wiping out climbers like a bowling ball in a perfect strike. There are deep crevasses that can swallow up a climbing team without their bodies ever being discovered. There are no trees on the mountain, nor any wood for campfires. Climbers have to share the mountain with bears and vicious wolf dogs, and poisonous snakes sometimes share your tent!

Several intrepid explorers and mountain climbers have climbed Ararat in modern times. Some walked away from the base as did Noah; others are still buried on the mountain. One who walked away was Dr. J. J. Friedrich Parrot who climbed the slope October 9, 1829. He was a recognized scholar and the Russian Imperial Councillor of State.

Major Robert Stuart, an Englishman, was successful in July of 1856. Douglas W. Freshfield, editor of the **Alpine Journal** came within 1000 feet of success, but was forced to turn back because of a terrible headache caused by the altitude. Christopher Trease climbed alone in 1965 without an ice axe and died there.

During the 1800s, the ark was seen by many local explorers including numerous Turkish military officials. The **Chicago Tribune** reported on the ark in its August 10, 1883 issue: "A paper at Constantinople announces the discovery of Noah's Ark. It appears that some Turkish commissioners appointed to investigate the question of avalanches on Mt. Ararat suddenly came upon a gigantic structure of very dark wood protruding from a glacier....There was an Englishman among them who had presumably read his Bible, and he saw that it was made of the ancient gopherwood of Scripture, which, as everyone knows, grows only on the plains of the Euphrates.

"Effecting an entrance into the structure, which was painted brown, they found that the admiralty requirements for the conveyance of horses had been carried out, and the interior was divided into partitions fifteen feet high. Into three of these only could they get, the others being full of ice, and how far the Ark, extended into the glacier they could not tell. If, however, on being uncovered, it turns out to be 300 cubits long, it will go hard with unbelievers."[12]  Of that, you can be **sure**.

In 1955, an expedition recovered wood from the ark almost 35 feet below the surface of an ice pack, and the wood revealed an age range from 1,200 to 5,000 years old. Early in the 1970s, American spy planes and military satellites photographed the wood structure on the Turkish mountainside.

We do know for sure that Noah and his family escaped the universal Flood by being obedient to God's command to build a large boat, and we know that the boat landed on Mount Ararat in Turkey. Why is it so incredible that the ark has been preserved in the ice until the "right" time for its  discovery? We don't need the ark to convince us that the Bible is the Word of God, but unbelieving men often need to be "jarred" into reality. Maybe, just maybe, Noah's ark will do that in the closing days before Christ returns for His own.

# CHAPTER FOURTEEN

# IS THE GEOLOGIC COLUMN TRUE?

Creationists believe that the geologic column was a direct result of the universal Flood and even unbelieving geologists, for hundreds of years, supported that position until Darwin fouled the scientific nest. Now, many geologists tell us that the geologic column has been developing for hundreds of millions of years, building up to the level we see when a hill is cut away for a road or housing development. The geologic column is what scientists call the layers of sedimentary rock that are sometimes thousands of feet thick.

According to their unproved (and unprovable) theory, the strata of sedimentation were the result of very slow erosion over the past "billions of years." We are told that streams carried dirt, dust, rocks and animal remains down to the seas where they were laid out horizontally on the ocean bottom. After long ages, during which the ocean beds were layered with stratum, those strata rose from the water to become dry lands while the ocean waters (that had previously covered the strata) flooded other continents that slowly sank beneath the water. Those areas would themselves be raised in the ages to come. Those strata make up the geologic column according to the myth makers.

Evolutionists make much of the geologic column where, the story goes, the "fact" of evolution is plainly seen.

Simple marine creatures are found on the bottom (where, it is assumed, the oldest rocks are found), then land plants and more complex creatures are found farther up the column in "recent" rocks. As you go up the column, (that allegedly reveals billions of years of time), the organisms get more complex. After all, the column is supposed to reveal the beginning of life and its progress finally to man. What the rocks of the column actually reveal is that evolutionists have rocks where most people have brains!

Most geologists take the position that the condition of the Earth is a result of normal, recognized physical laws that have operated in a uniform way throughout time (uniformitarianism). Creationists believe that the condition of the Earth is clearly the result of a world-wide catastrophic Flood (catastrophism).

Scientists have arranged time into geologic ages, represented by different portions of the sedimentary strata found over most of the Earth; however, they never tell students that geologists admit, "There is no place on earth where a complete record of the rocks is present."[1] The complete geologic column only exists in textbooks!

The main eras are called Proterozoic which means, "before life," Paleozoic, meaning "ancient life," Mesozoic, meaning "intermediate life," and Cenozoic, meaning "recent life." The different eras are further divided into "periods" and "epochs."

## DATING FOSSILS

The Grand Canyon reveals more of the geologic column than any other place on Earth, but about 345 million years are missing! It has multi-layers of sedimentary rock, sometimes thousands of feet thick, loaded with fossils. That impresses students, but they aren't told that out-of-place fossils are found throughout the various stratum. Evolutionists never talk about that unless forced by pushy creationists!

How do they know how old a stratum of rock is? Well, by the fossils found in the stratum. The follow-up question is, "But, how do they know how old the fossils are?" We are told that since evolution is a fact, the simple creatures must belong on the bottom, at the very beginning of life. If remains of an ape are found, then it belongs near the top because apes evolved in recent times. That is **not** science; it is philosophy.

When many fossils of the same creature are found in a stratum, those fossils are called, "index" fossils. What happens if index fossils turn up **alive** such as coelacanth, the "link fossil" that was caught off the coast of Madagascar in 1938? Well, it makes for a bad day for evolutionists because coelacanth allegedly disappeared seventy million years ago according to them. It means that any rock that was dated according to the coelacanth index is now as young as 100 years old![2] What a kick in the teeth for evolutionists!

Geologists date the rock strata not by the physical or chemical composition, but by the embedded fossils found within a stratum. However, the only way one can know the age of a fossil is if it lived **only** during a certain time in history. This would mean that there must have been different creatures living at different times in history resulting in a definite index to the chronology.[3]

The question arises, "How do we know what was

living when?" Ah! Evolution gives us the answer: All creatures developed from simple to higher creatures, so it is presumed that when many horse fossils and a trilobite are found in the same stratum, the stratum is young. It is young (recent) because the horse "evolved" only recently while the trilobite was at the very bottom of animal evolution.

But what is the trilobite doing with horses since they existed millions of years earlier? **That** is a major problem for evolutionists, but they have developed an answer: The out-of-place fossils are called, "stratigraphic leaks" and "re-worked fossils." Those fossils found too high in the column are called, "re-worked" fossils, and those too low are known as "stratigraphic leaks." The additional problem with out-of-place lower fossils is that they are found in stratum formed **before they lived**! Good trick if you can do it!

## OVERTHRUSTING

Often thick strata are found to be in the "wrong" order. Strata that have an abundance of a particular fossil are identified as belonging to the era when that creature lived (according to evolutionists). Those fossils, found in abundance, are known as "index fossils." But if strata are found near the top of the column with index fossils from the **bottom** of the column, geologists must explain how that can happen. It's called "overthrusting." A good example is in Glacier National Park where a block of Pre-Cambrian limestone that is supposed to be a billion years old has been found on **top** of Cretaceous shale that is "only" 100 million years old![4]

We are expected to believe that a block of limestone

300 miles long, 25 miles wide, and over 2 miles **thick,** moved about 30 miles horizontally over the "younger" area by various unknown forces, and now the "older" sits on top of the "younger" formation! If that could be proved, and it can't, then it would demolish uniformitarianism because that geological activity is not taking place anywhere in the world today. And evolutionists tell us the Flood is an outrageous teaching!

Dr. Clifford Burdick, consulting geologist, spoke very pointedly of the above overthrust: "In order to get them (the fossils) in the right order, evolutionists propose that a block of the earth's crust 300 miles long and 25 miles wide and 2 miles in thickness, was turned upside-down...If there was such movement there are three criteria to look for. One is there should be ground-up rock between the two layers, like the upper and nether millstone...Another criterion is breccia, i.e. broken up rock fragments...Then there are slickensides (gouge marks)...We spent years in efforts to find such evidences at the contact points in Glacier Park, and **we have yet to find one single case of contact**. We are forced to say that the rocks in Glacier Park are in the **same order** in which they were laid down...not millions of years ago, but thousands of years ago."[5] (Emphasis added.)

The geologist is saying that the physical evidence is not there for overthrusting. The massive layers of rock did not move "out of place." He went on to say that the rock layers do not have the inner strength to withstand the massive pressure needed to push billions of tons of solid rock over the lower rocks. He says, "...it is incredible to me that evolutionists will hang on to the idea of thrust fault just to save evolution concepts."[6]

Only simple-minded folk would believe that huge sheets of rock, over 2 miles thick, could have moved across other rock surfaces for many miles. Scientists writing in the **Bulletin of Geological Society of America** admit, "Since their earliest recognition, the existence of large overthrusts has presented a mechanical paradox that has never been satisfactorily resolved."[7] While small areas are known to have experienced overthrusting, the areas always have physical evidence of that fact.

The Matterhorn, a famous Swiss mountain, is out of place according to evolutionists! Yes, it has moved from thirty to sixty miles over younger rock formations to its present location! But they get more desperate, for evolutionists tell us the famous Mythen Peak of the Alps has been pushed (by whom or what?) from Africa to Switzerland![8] (Hey, don't look at me like that. **I'm** not the one who is wacko. Go talk to your local evolutionist. Ask him for a reasonable answer, and watch him squirm. Evolutionists don't squirm with grace.)

Scientists stoop to such absurdities because they have "bought into" the geologic column that is supposed to tell the story of geologic history from the beginning of time to now, but even some scientists have had problems accepting their own conclusions.

## SIMPLE TRUTH

W. W. Watts, who was president of the geology section of the British Association for the Advancement of Science, said that the explanations given are "...so hopelessly inadequate that we sometimes feel compelled to doubt whether the facts are really as stated," then adds, "But the phenomena have now been so carefully observed and in so many districts, that any real doubt as to the facts is out of the question."[9]

My, my, what lengths unbelieving men will go to, trying to explain away God working in man's affairs. Why posit absurd theories when the simple truth is so much more acceptable? Why not accept as truth what seems to be reasonable? The Flood of Genesis deposited fossils promiscuously without regard to their simplicity or complexity. Moving water, carrying sand, dirt, shells, fossils, etc. moving back and forth, sometimes fast, sometimes slow, over many days would drop the material in horizontal layers one on top of the other exactly as we see in the strata all about us.

The geologic time periods are supposed to represent Earth's geologic history back to the very beginning. What evolutionists never tell you is that the column is a paper column! It doesn't exist in reality! The column is alleged to be 150 miles in depth, but that is "putting columns together" from all over the world.

The actual column (what there is of it) is made up of sedimentary rocks that were formed when running water, carrying dirt, dust, and other residue, slowed down dropping the heavier elements to the bottom of the stream. Then, through pressure, the material was cemented together to produce sedimentary rock.

Often, the remains of some creature were caught in the stream, carried along with the sediment, and dropped along with the sediment when the water flow slowed. Buried in the sediment, under pressure, the creature's bones become part of the rock--a fossil.

Evolutionists tell us that sediment was laid a grain at a time, while creationists believe it all happened during the Flood. Three-fourths of the world is covered by sedimentary strata, so it is obvious to the most dull person that water played the major role in the geologic conditions of the Earth.

Evolutionists assume that all life evolved from very simple to more complex creatures, so each evolving creature would leave its trace in the strata or column. As a creature evolved, died, and was deposited on the bottom of some stream, it became part of some stratum, and that creature (fossil) would be found someday and classified as old or young by some eager geologist.

## CIRCULAR REASONING

"But," says the naive one, "how do you **know** when a creature lived?" The exasperated evolutionist tells him that life started with the most simple creatures, and got progressively more complex, so obviously the simple creatures will be at the bottom strata and the more complex will be found as one goes up the geologic column.

So one must assume that evolution is a fact in order to date the fossils in the column that prove the fact of evolution! Huh? In college we called that circular reasoning. It proves nothing except that evolutionists are going around in circles. The main proof of evolution rests upon the assumption of evolution! Well, if this is science, then so is the story of "The Three Little Pigs"! In fact, it makes more sense than does evolution.

This circular reasoning was done by Darwin when he

coined the term, "survival of the fittest" to prove evolution. When some poor soul asks how you can identify the fittest, he is told, "all those who survive"!

In a 1979 interview with Dr. Donald Fisher, the state paleontologist for New York, Luther Sunderland asked him how he dated fossils. Answer: "By the Cambrian rocks in which they were found." When asked if this was not circular reasoning, he replied, "Of course; how else are you going to do it?"[10]  Out of the horse's mouth!

Even the **Encyclopaedia Britannica** admitted circular reasoning when they wrote, "It cannot be denied that from a strictly philosophical standpoint geologists are here arguing in a circle. The succession of organisms has been determined by a study of their remains embedded in the rocks, and the relative ages of the rocks are determined by the remains of organisms that they contain."[11] So the blind lead the blind, and they all fall into the ditch!

Dr. Walter Brown reports in his book, **In the Beginning**, on spores of ferns and pollen from flowering plants that were found in Grand Canyon rocks allegedly deposited before life evolved (according to evolution)! A leading authority on the Grand Canyon photographed horse-like hoof prints that **predate horses and other quadrupeds** by a hundred million years,  according to Brown.[12]

Can you understand why evolutionists get super-sensitive when a creature is found "in the wrong" stratum? Their outrageous attempts to explain it are often hilarious. There are many indications that man and various animals lived contemporaneously from the beginning and did not evolve over millions of years. This fact alone wipes out **all** kinds of evolution.

## TRILOBITES AND DINOSAURS

There are places in Zimbabwe and Arizona where pictographs have been found on cave and canyon walls that show dinosaurs, drawn by "early" man. Is it possible, gasp, do I dare say it, that man and dinosaurs lived at the same time? Yes, this is true, my friend. Human prints and dinosaur prints have been discovered in the same stratum in New Mexico, Arizona, Mexico, Kentucky, Missouri, Russia, Illinois, Texas and other locations!

Evolutionists D. H. Milne and S. Schafersman wrote in the **Journal of Geological Education** about the consequences of finding human prints and dinosaur prints together: "Such an occurrence would seriously disrupt conventional interpretations of the biological and geological history and would support the doctrines of creationism and catastrophism."[13] They are right on target.

Glen Rose, Texas, well-known as dinosaur country, is where Dr. Carl Baugh has been excavating for many years. I have participated in a very limited way in his work. The area is known to have been rife with dinosaurs about seventy million years ago or five thousand years ago, depending upon whether one is an evolutionist or a creationist.

Dr. Baugh wrote a thoroughly exciting and delightful book on his work, but I will only deal with that part which interests me at the present. Human and dinosaur prints found at Glen Rose have been documented for many years, but Dr. Baugh has done the most recent work in the area.

On March 16, 1982, Dr. Baugh, accompanied by skeptics, removed a layer of limestone twelve inches thick to reveal human and dinosaur prints within inches of each other! By the next day, they had uncovered four human footprints and twenty-three dinosaur prints![14] The facts are in: humans and dinosaurs shared the same environment at the same time, and that fact blows the geologic column and evolution out of the water!

Not only did men and dinosaurs live together, men and trilobites were neighbors! Yes, I know that trilobites disappeared 230 million years ago according to the hysterical evolutionists, but it isn't so.

On June 1, 1968, William Meister, found the fossils of several trilobites in Utah in what appeared to be the fossilized, sandaled footprint of a man![15] That could not happen if the geologic column is correct.

Norm Sharbaugh reports of another human print in his excellent book on evolution. In Pershing County, Nevada, on January 27, 1927, Albert Knapp found a shoe print in Triassic limestone (dated 225 million years by evolutionists). The shoe's double row of stitches could be clearly seen in the micro-photograph and were more refined than shoemakers were using in 1927![16] Do I need to ask how a human print could be found in limestone millions of years before man existed (according to evolutionists)?

Samuel Hubbard of the Oakland Museum in California said of the above print, "There are whole races of primitive men on earth today, utterly incapable of sewing that moccasin. What becomes of the Darwinian theory in the face of this evidence that there were intelligent men on earth millions of years before apes were supposed to have evolved?"[17] Great question! How about some answers from Gould, Eldredge and Sagan?

Dr. Clifford Burdick, a geologist, found evidence that men and trilobites did live together when he discovered the footprints of a barefoot child, one of which contained a compressed trilobite![18] Some days it doesn't pay for an evolutionist to get out of bed!

## CAMBRIAN EXPLOSION

But they have other problems with the column such as how life appears abruptly in the Cambrian strata, the very lowest level in the column! The lower four-fifths of the rock of the earth's crust is without **any** signs of life! Then, all at once, life abruptly appears out of nowhere! Maybe, as if it were created?

Paleoanthropologist Richard Leakey, who is an evolutionist, made a cogent statement on this problem for evolutionists: "Geologists of the 1830s were convinced that they had discovered the dawn of life in this strata, called the Cambrian, and no hypothesis save special creation was deemed adequate to account for the sudden appearance of Cambrian fossils. Rocks older than the Cambrian were, so it seemed, completely devoid of fossils. In the second place, the various strata each generally had its own characteristic fossil flora and fauna and the transitions between strata were abrupt. This suggested to geologists the probability of successive wholesale creations and extinctions...."[19] And the spinners of the fairy tale have no answers.

So the evolutionists have a big problem. If everything evolved over billions of years, there would be some evidence, and while there are fossils from Cambrian (and higher), there is nothing below that could possibly have been ancestral to the Cambrian creatures. Where did the Cambrian fossils come from? What happened to their ancestors? With evolution you must have ancestors; with creation they are not necessary.

D. Axelrod, a top evolutionist, commented on this major question. "...when we turn to examine the Precambrian rocks for the forerunners of these Early Cambrian fossils, **they are nowhere to be found**. Many thick (over 5,000 feet) sections of sedimentary rock are now known to lie in unbroken succession below strata containing the earliest Cambrian fossils. These sediments apparently were suitable for the preservation of fossils because they often are identical with overlying rocks which are fossiliferous, yet **no fossils are found in them**."[20] (Emphasis added.)

Evolutionists don't explain this problem, because they **can't**. Rather than the earliest forms of life (that evolution demands) appearing in the Cambrian strata, representations of all but the vertebrates are found there! More than 5,000 species are found in the Cambrian layers; and no evolution is evident! None! What you do see are fossils (allegedly 380 million years old) that are **exactly** like our modern creatures. Where is the evolution?

Professor John W. Klotz wrote, "In what is known as the Cambrian period there is literally a sudden outburst of living things of great variety. Very few of the groups which we know today were not in existence at the time of the Cambrian period. One of the problems of this Cambrian outburst is the sudden appearance of all these forms. All the animal phyla are represented already in the Cambrian period except two minor soft-bodied phyla (which may have been present without leaving fossil evidence) and the chordates.

Even the chordates may have been present, since an object which looks like a fish scale has been discovered in Cambrian rock."[21]

Dr. Henry Morris adroitly comments on this problem in his classic book, **The Biblical Basis for Modern Science**: "Evolutionists used to claim that the reason for the sudden appearance of complex creatures in the Cambrian, with no fossil precursors, was that their ancestors were all soft-bodied and thus left no hard parts to be fossilized. As Stanley pointed out, however, many fossils of soft-bodied animals have been found in the Cambrian (and later) rocks, so there is no good reason why soft-bodied creatures could not have been fossilized in the Precambrian, if they existed. It is especially noteworthy that all of the great phyla [major animal groups] have been found in rocks of the Cambrian, supposedly the oldest of the fossil-bearing rock systems."[22] Where are the remains of their parents? Of course, they could not have existed before creation!

Evolutionists require very long periods of time to patch together their con job, but how can highly complex life appear all at once if it evolved over long periods of time? No fossils have been found leading up to those in the Cambrian Period. It seems as if, well, everything was created at the same time! There is not a **single** ancestor for any of those Cambrian fossils anywhere on the face of the Earth--or under the Earth! World renowned evolutionist, George G. Simpson has characterized the absence of Precambrian fossils as the "...major mystery of the history of life."[23]

An incredible admission was made by Richard Goldschmidt, well-known geneticist and evolutionist who said that "practically all orders or families known appear suddenly and without any apparent transitions."[24] This problem even gave Darwin "fits." He wrote in **Origin**, "The abrupt manner in which whole groups of species suddenly appear in certain formations, has been urged by several paleontologists...as a fatal objection to the belief in the transmutation of species. If numerous species, belonging to the same genera of families, have really started into life at once, the fact would be **fatal to the theory of evolution through natural selection**."[25] (Emphasis added.) Well, that's some admission from the Guru of Gradualism.

Following his admission in **Origin**, Darwin said: "during these vast periods the world swarmed with living creatures....To the question why we do not find rich fossilferous deposits belonging to these assumed earliest periods prior to the Cambrian system, I can give no satisfactory answer....The case at present must remain inexplicable; and may be truly urged as a valid argument against the views here entertained...."[26] And, after more than 100 years, there's still no "satisfactory" answer, and the case still remains "inexplicable." And, yes, I'm using it as a "valid argument" against his views!

Niles Eldredge, a paleontologist at the American Museum of Natural History wrote of the abrupt appearance of "the earliest known representatives of the major kinds of animals" in the Cambrian period and was cognizant that creationists have made much to do about it. He admitted that it "does pose a fascinating intellectual challenge."[27] He added that the Cambrian evolutionary explosion is still shrouded in mystery.[28] Well, not if you believe the Bible, and if you accept the scientific facts. The fact is that life appeared **suddenly** when God created everything.

## ABSENCE OF TRANSITIONAL FOSSILS

Another major problem for evolutionists is similar to the absence of Precambrian fossils: Why are there no transitional fossils in the geologic column? If life has been slowly evolving, there should be **billions** of fossils in the in-between stages, but to the embarrassment of the Disciples of Darwin, there are none!

As to the gap between invertebrates and vertebrates, H. Smith of New York University confessed, "The gap remains unbridged, and the best place to start the evolution of the vertebrates is in the imagination."[29] Yes sir, in the imagination! That is great science!

Dr. Colin Patterson, Senior Paleontologist, British Natural History Museum, further sustains our position: "Gould and the American Museum people are hard to contradict when they say that there are no transitional fossils....I will lay it on the line--there is **not one** such fossil for which one could make a watertight argument."[30] Not one!

But read what Dr. David Kitts said: "The fossil record doesn't even provide any evidence in support of Darwinian theory."[31] That's from an evolutionist!

World famous evolutionist, Dr. Niles Eldredge wrote, "The search for missing links is probably fruitless...**no one has yet found any evidence of transitional forms**."[32] (Emphasis added.) Eldredge admits that they have never found **one** in-between fossil after digging around in the geologic column for over a hundred years! Question: Why is gradualism still being taught in the public schools?

Mr. Darwin's famous confession is applicable here:

"...why is not every geological formation and every stratum full of such intermediate links? Geology assuredly does not reveal any such finely graduated organic chain; and this perhaps is the most obvious and gravest objection which can be urged against my theory."[33]

## ABSENCE OF METEORITES

Another major problem for rabid evolutionists is the absence of meteorites in the geologic column. Meteorite showers have been occurring throughout history, yet W. H. Twenhofel honestly admits that, "No meteorites have ever been found in the geologic column."[34] Other experts agree with him.

W. A. Tarr concurred, "I have interviewed the late Dr. G. P. Merrill, of the U.S. National Museum, and Dr. G. T. Prior, of the British Natural History Museum, both well-known students of meteorites, and neither man knew of a single occurrence of a meteorite in sedimentary rocks."[35] And Stansfield said, "No meteorites have been found in the geological column."[36]

Understand the problem: We are told that the geologic column was laid down over billions of years so if meteorites fell to earth, they would have to be found in the column. However, creationists maintain that the "column" was formed rather quickly, and there are no meteorites there because none fell during the short period of time following the Flood when the strata were laid down as the flood waters receded.

Evolutionists wish that the absence of meteorites were their only problem, but they have many.

## MORE PROBLEMS

Evolutionists believe that the column took billions of years to form, and the ages can be "read" like a book, but it isn't so. In addition to the difficulties already mentioned, there is the problem of trees piercing through successive layers of stratum. How could a tree lay for millions of years, and be covered very slowly by sediment without rotting? That does not happen in the real world.

There have been man-made artifacts found in the column, in solid rock, such as a thimble, bell, gold chain, spoon, metal vase, screw, nails, coin and a doll. If the rock strata were laid over billions of years ago, how could human objects be embedded into the rock **before** they were made?

An iron hammer, encased in stone, was discovered near London, Texas in strata about four hundred million years old--according to how evolutionists compute time. I have examined that hammer twice, since it is owned by my friend, Dr. Carl Baugh. How could a man-made object be four hundred million years old? Of course, even evolutionists know that could not be true.

Yes, the foundations of evolution are cracking as can be seen by the above evidence and by statements like that by Dr. E. J. H. Corner, Professor at Cambridge University: "...I still think that to the unprejudiced, the fossil record of plants is in favor of special creation."[37]

I'm convinced that most evolutionists are not unprejudiced. They must seek to explain every phenomenon of nature from a humanist perspective whether or not the evidence supports it. Whenever their research brings them even close to a Creator, they flee in the other direction as quickly as a mythical vampire flees the sunlight. Or is that Sonlight?

Ellen Boys Overlooking the Grand Canyon

# CHAPTER FIFTEEN

# IS UNIFORMITARIANISM SCIENTIFIC OR BIBLICAL?

The geologic column and fossil record are interpreted by a system called "uniformitarianism." A standard textbook on the subject of historical geology affirms, "The up-rooting of such fantastic beliefs [that is, those of the catastrophists; creationists] began with the Scottish geologist, James Hutton, whose **Theory of the Earth**, published in 1785, maintained that **the present is the key to the past**, and that, given sufficient time, processes now at work could account for all the geologic features of the Globe. This philosophy, which came to be known as the **doctrine of uniformitarianism** demands an immensity of time; it has now gained universal acceptance among intelligent and informed people."[1] (Emphasis his.)

That is, "intelligent and informed" people who are looking for **any** answer that excludes God. Uniformitarianism is not science since it cannot be proved. It is a philosophy, a **belief** in a system of thought. Or, as Hutton himself called it, it is a "doctrine."

We are all aware that there is a form of uniformitarianism (or, more correctly, a principle of uniformity), in that God's laws are at work, and His laws and the universe are completely compatible. There is uniformity

214 Evolution: Fact, Fraud, or Faith?

in the stars, moon, sun, rotation of the Earth, etc.--all set in motion by God, but evolutionists insist that all things are working today as they have for "billions of years." And they are working at the same rate. Of course, that eliminates creation, the Flood, etc., permitting only local events that might have only a short term impact on an area. One thing is sure for the uniformitarians: world catastrophic events are **not** part of their system because everything is as it has always been. The Bible predicted this false teaching about two thousand years before it was hatched in the confused mind of Hutton.

## BIBLICAL

The Apostle Peter described these days and the above theory perfectly in II Peter 3:3-6 where he wrote, "Knowing this first, that there shall come in the last days scoffers, walking after their own lusts, And saying, where is the promise of his coming? for since the fathers fell asleep, **all things continue as they were from the beginning** of the creation. For this they are willingly ignorant of, that by the word of God the heavens were of old, and the earth standing out of the water and in the water: Whereby the world that then was, being overflowed with water, perished."

Well, that does describe our day, especially as it relates to the cult of uniformitarianism. Notice that Peter makes it clear that the world **did** perish by the Flood, and he warns us that men would arise affirming that "all things continue as they were from the beginning." God, they say, has not intervened in the affairs of men. God did not send a flood to destroy the world. They are wrong!

God did not lead the Jews out of Egypt with the attending plagues. God did not part the Red Sea to give safe passage to the Jews and to destroy the pursuing Egyptian army. God did not interfere with the passage of the sun for Joshua, and of course, Christ was not raised from the dead! None of those things could have happened, because everything continues today, according to natural laws, as they have in the past. Wrong again!

This false system of uniformitarianism was popularized by Charles Lyell, a lawyer, who died in 1875. He held that the earth's changes were gradual, taking place at the same uniform rate of slowness that they are today. Lyell is thus credited with the propagation of the premise which more or less has guided geological thought ever since, namely that the **present is the key to the past**. In essence, Lyell's doctrine of uniformitarianism states that past geological processes operated in the same manner and at the same rate as they do today. God is nowhere to be found, nor is He needed. And neither is He wanted.

The key to Lyell's false teaching is "natural." Every effect in the past had a natural cause similar to those causes seen in nature today. Geisler calls such uniformitarianism not science, but "scientism." He added, ""It is not the scientific study of nature; it is philosophical naturalism."[2] That is **exactly** what it is.

Most scientists approach the issue with a built-in prejudice. Their philosophy (read: religion) will not permit them to consider that they have been worshipping at a false (and pagan) altar, and that the Flood is the answer to the geologic conditions of the world today.

However, one honest, renowned, British paleontologist, L. M. Davies, made an unusual admission on this subject: "Here, then, we come face to face with a circumstance which cannot be ignored in dealing with this subject...namely, the existence of a marked **prejudice** against the acceptance of belief in a cataclysm like the Deluge. Now we should remember that, up to a hundred years ago, such a prejudice did not exist...as a general one, at least. Belief in the Deluge of Noah was axiomatic, not only in the Church itself (both Catholic and Protestant) but in the scientific world as well. And yet the Bible stood committed to the prophecy that, in what it calls the 'last days,' a very different philosophy would be found in the ascendent; a philosophy which would lead men to regard belief in the Flood with disfavor, and treat it as disproved, declaring that 'All things continue as from the beginning of the creation' (2 Peter 3:3-6). In other words, a doctrine of Uniformity in all things (a doctrine which the apostle obviously regarded as untrue to fact) was to replace belief in such cataclysms as the Deluge."[3] (Emphasis his.)

That scientist talks like a Christian; at least he has been exposed to Bible principles, and is not antagonistic toward them. He is saying that, until Darwin, most scientists accepted the Flood as the explanation for the world's geologic condition. Then, there was a shift in **philosophy** that colored their approach to science. Of course, the geology hadn't changed, but one's perception of geology (and other divisions of science) had now changed. God was no longer necessary. Since creation was out, there had to be a replacement for it, and that replacement was evolution. But evolution required immense ages. Uniformitarianism gave them the time needed to spin their fairy tale.

Evolutionist Dr. George Wald wrote in **The Origin of Life**, "Time is in fact the hero of the plot...Given so much time, the impossible becomes possible, the possible becomes probable, and the probable virtually certain. One has only to wait; time itself performs miracles."[4] (Wald has recently backed away from this position.)

Of course, secular "miracles" are acceptable, but divine miracles are anathema!

## SCIENTIFIC

So uniformitarianism, masquerading as science, demands that the present Earth processes can explain the very gradual changes and development of **everything** over many billions of years. We are told, and are expected to believe, that, with enough time, anything can happen--even secular miracles. So, the stars, planets, space, rocks, flowers, bugs, snakes, birds, apes and man all developed over billions of years.

Of course, a high school student knows that the Second Law of Thermodynamics teaches that the universe is running down (while evolution requires it to run uphill) and long periods of time will only add to the decay of the universe.

The Second Law is universally true, although no one knows exactly why it is true. It has always been found to be true whenever it has been tested, but there is no natural explanation.

Ross, of Harvard University, tells us that the Second Law is always consistent, and he puts to rest the evolutionist argument about open and closed systems: "...there are no known violations of the second law of thermodynamics. Ordinarily the second law is stated for isolated systems, but the second law applies equally well to open systems. However, there is somehow associated with the field of far-from-equilibrium phenomenon the notion that the second law of thermodynamics fails for such systems. It is important to make sure that this error does not perpetuate itself."[5]

God wound up the universe at the creation, and it has been running down ever since, not running up! Some desperate evolutionists have retreated to the position that the universe has always existed (that gets away from **any** explanation as to origins), but astronomer Robert Jastrow answered that the "laws of thermodynamics...indicated that the Universe had a beginning."[6]

The running down process is a general law of nature accepted by all scientists. As the years pass, we all get older and weaker. Your car rusts, tires wear out, the engine finally quits.

But evolution is at odds with that second law, and astronomer Sir Arthur Eddington admonished evolutionists, "If your theory is found to be against the 2nd Law of Thermodynamics I can give you no hope."[7] Isn't it a major tragedy when people are without hope? What a miserable life that must be, but evolutionists have no hope in this world or the world to come!

Evolution says that man started in a primordial soup billions of years ago (where did the "soup" come from?) and has been climbing up ever since, all contrary to the universal Second Law of Thermodynamics. Christians would say that man has not risen but fallen!

The Second Law is clearly taught by Isaiah in chapter

51, verse 6: "Lift up your eyes to the heavens, and look upon the earth beneath: for the heavens shall vanish away like smoke, and the earth shall wax old like a garment, and they that dwell therein shall die in like manner..."

In Matthew 24:35 Christ reminds us, "Heaven and earth shall pass away, but my words shall not pass away." The Old and New Testaments teach, in many places, the Second Law. The Earth is running down, not up. And man didn't fall up, but down!

Yes, there are universal changes taking place, but those changes are of degeneration, decay, disintegration and death as supported by the Second Law and in opposition to uniformitarianism and evolution. Since they are opposites, and contradict each other, they are not both correct. One is wrong, and that is uniformitarianism.

This fact has been admitted by top evolutionists in recent years! Evolutionist Stephen J. Gould wrote: "Substantive uniformitarianism as a descriptive theory has not withstood the test of new data and can no longer be maintained in any strict manner."[8]

Gould, no friend of creationists, bashed Lyell's uniformitarianism when he concluded, "...Lyell's 'uniformity' is a hodgepodge of claims. One is a methodological statement that must be accepted by any scientist, catastrophist and uniformitarian alike. Other claims are substantive notions that have since **been tested and abandoned**. Lyell gave them a common name and pulled a consummate fast one: he tried to slip the substantive claim by with an argument that the methodological proposition had to be accepted, lest 'we see the ancient spirit of speculation revived, and a desire manifested to cut, rather than patiently to untie, the Gordian knot."[9] It would seem that evolutionists do a better job of cutting up their own than creationists do!

Evolutionists quickly accepted the philosophy of uniformitarianism because they needed to supply the long ages, and it was "natural"--not requiring anything supernatural. So they jumped on the bandwagon, but are now abandoning it one by one.

George James Shea wrote that Lyell's uniformitarianism, "specifically his ideas on identity of ancient and modern causes, gradualism, and constancy of rate, has been explicitly refuted by the definitive modern sources as well as by an overwhelming preponderance of evidence that, as substantive theories, his ideas on these matters were simply wrong."[10] He later commented that, "The idea that the rates or intensities of geological processes have been constant is so obviously contrary to the evidence that one can only wonder at its persistence."[11]

So scientists are slowly backing off from uniformitarianism, and a few are embracing creationism because they realize that the geologic formations of the Earth indicate some massive catastrophes have taken place that seem to defy, dispute and disprove uniformitarianism. This seems to be what evolutionist Niles Eldredge is saying in his book, **The Monkey Business:**

"For well over a century now, we have spoken of Ice Ages in the recent geological past...Volcanoes and earthquakes are truly sudden in their action, and many wreak huge changes. And the recent invocation of an asteroid impact as the trigger to the ecological collapse that ended the Cretaceous world and wiped out perhaps as many as ninety percent of all living species is a catastrophe **par excellence** (and even if there was no asteroid, the extinction 'event' itself was a disaster)."[12]

## HISTORIC CATASTROPHIC EVENTS

We have been reading for years about various postulates that seek to explain how the dinosaurs were wiped off the face of the Earth (an incident that would have had to be global), and other catastrophic events in the past or prophesied for the future--all in opposition to uniformitarianism. Following are a few events, so catastrophic in their results that they could hardly be characterized as "normal" or meet the criterion of the uniformitarians.

In 1887, a China flood took the lives of 900,000 people, and in 1931 China experienced another incredible deluge that swept 3.7 million souls into eternity. Hardly the norm!

Earthquakes, volcanoes and other physical activity have changed the face of the Earth, all in defiance of uniformitarianism. Earthquakes have reshaped the Earth, and killed millions of people. In 526 A.D., over 250,000 people were killed in Antioch, Syria while in 856 A.D. at least 200,000 people were killed in Iran. In 1201 A.D., a million people died in the quake in upper Egypt. In 1556 A.D., 830,000 people died in a China quake while in 1976, 240,000 more died in another quake.

One of the worst storms to strike the Earth occurred October 7, 1737 near Calcutta, India. A 39-foot wall of water surged through the villages and cities killing 300,000 people! In 1881, at Haiphong, Vietnam, almost as many people were killed when a typhoon hit.

The Earth probably changes more from volcanic eruptions than from any other natural activity. At about 4600 B.C., Mount Mazama (12,000 feet high) exploded, and lost 5,000 feet immediately! It threw seventeen cubic miles of rock into the sky. Then the core collapsed and water drained in to form Crater Lake to the depths of 1,932 feet![13]

A spectacular eruption occurred in August of 1883, in what is now Indonesia. There was a "burst with a roar heard more than 2,000 miles away; the resultant aerial vibrations circled the globe several times....Volcanic dust blocked out the sun for a radius of 100 miles....Finally came the culminating horror--the tidal waves. More than 100 feet high and with a velocity in excess of 50 miles an hour; they ravaged the nearby coasts of Java and Sumatra....This disaster crushed the life from more than 36,000 people."[14] Hardly uniformitarian!

Most of us were witnesses to one of the most incredible non-uniformitarian examples in history--Mount St. Helens. It was on May 18, 1980 at 8:32 a.m. that the Washington state mountain's summit exploded along with the north slope. The initial resultant explosion was equivalent to 20,000 Hiroshima-size bombs! That initial eruption ripped up 150 square miles of forest in six minutes, and within six years strata up to 600 feet thick had formed!

Then on June 12, 1980, a 25 feet thick deposit accumulated, proving that evolutionary dogma is not valid: strata do not have to be laid slowly over long periods of time, but can take place in days--as creationists have been saying for years. Only five days after the eruption, the rills and valleys resembled the Badlands topography! A new canyon, 100 feet deep, was formed, gouged out of solid rock! The north fork of the Toutle River was blocked by a landslide, and on March 19, 1982, it was unblocked by a mud flow. One day later, five canyons (that converge) with clefts over 100 feet high had been gouged out proving once again that it doesn't take million of years of erosion to produce a canyon.[15] Surely the most rabid evolutionist doesn't call that uniformitarian activity.

An awesome, earth-changing quake occurred in the small town of New Madrid, Missouri, about 100 miles south of St. Louis, on December 15, 1811. Buildings came crashing down, and fissures opened in the earth as buildings swayed hundreds of miles away from the center of the quake. As frantic people ran into the streets, they saw the land moving like ocean waves! At daybreak, the land was still churning, ripping down forests, swallowing hills and scooping out lakes, one 70 miles long!

The potent quake made itself known all over North America with church bells ringing in Virginia, chimneys falling in many eastern states, and terrifying Indians in northern Canada. For a while the Mississippi River was lifted up and flowed backwards! Then the backed up water came crashing downstream. Violent quakes were felt for months after the initial shock.

Creationists constantly repeat the fact that the universal Flood with its attendant earthquakes, seaquakes, volcanoes, violent winds, tidal waves and constant shifting of water was the cause of the world's geological condition. More and more, sane, sensible people are coming to the same conclusion. If an honest person knew **nothing** about the Bible or about evolution, he would be led to believe that a major catastrophe had produced the world's geology.

## PAST IS THE KEY

The September, 1977 issue of the **Reader's Digest** shook up the soft-headed sophisticates with this heading, "There is now compelling evidence to support the ancient accounts that there was indeed a vast, universal Deluge."[16] Boy, that caught their eyes!

The article continued, "Most scientists have agreed that it [The Flood] was based upon a local event...magnified in the retelling, gilded as a legend and finally concretized into the heroic stature of myth. Now, however, there is evidence that seems to indicate the Deluge might indeed have taken place. This startling conclusion comes, as we shall see, from the independent probing of two separate disciplines: geologists who have studied the shells of a tiny sea creature which lived when it happened 11,600 years ago, and archaeologists who have deciphered what was written by the hand of man some 8,000 years ago."[17]

Yet, most evolutionists preach, with the fervor of a hungry evangelist, the fantasy of uniformitarianism as a fact. Honest scientists know it is a fraud.

The disciples of uniformitarianism cannot explain why fossils are no longer being formed; why animals are no longer being quick frozen; why coal and oil are no longer being formed; and why major overthrusts are not taking place. All these things **must** continue if uniformitarianism is correct, but then they are not, and evolution loses one of its main props.

Even though the evidence against this theory is devastating, desperate men still hold to this pitiful philosophical ploy like a whimpering child clings to his security blanket. How pathetic to see educated adults take pure speculation and call it science because they will not admit the possibility of a sovereign God. We Christian creationists know that the present is not the key to the past but the **past** (creation, Fall of man, death and resurrection of Christ, the Word of God, etc.) is the key to **both** the present and the future.

# CHAPTER SIXTEEN

# DOES LIFE COME FROM NON-LIFE?

Science and the Bible both teach creation, corruption, and catastrophe. But evolutionists, doing their best to stay far from reality, have accepted absurdities like spontaneous generation to prop up their story. Their major problem (and they have **many**) is to get life started without any supernatural involvement. They still have that problem, but they get around it by **pretending** they don't have the problem. Very bright, these boys!

A scientist, who must remain anonymous, warned us to be suspicious of those who seek to explain the origin of life. He said, "Those who work on the origin of life must necessarily make bricks without very much straw, which goes a long way to explain why this field of study is so often regarded with deep suspicion. Speculation is bound to be rife, and it has also frequently been wild. Some attempts to account for the origin of life on Earth, however ingenious, have shared much with imaginative literature and little with theoretical inference of the kind which can be confronted with observational evidence of some kind or another."[1]

Evolutionists teach that life has progressed very slowly from non-life to modern man. Of course, they can't prove that position, but that has never bothered the evolutionists. For centuries, beyond Aristotle, men believed that life arose "spontaneously." It was an accepted "fact" (like evolution today?) that insects arose from mud and slime and pond water produced frogs. Rats came from garbage! So, of course, why couldn't man have started in a similar way?

Then in the 17th Century, an Italian physician named Francesco Redi proved, through controlled experiments, that spontaneous generation was not a "fact," but a fallacy. He proved that maggots, in rotting meat, arose only by reproduction of like living things. His work was soon forgotten.

Then in 1860, Louis Pasteur, a creationist, again exploded the myth of spontaneous generation through his work with bacteria. Now the "beast" should stay dead: spontaneous generation was impossible, could not happen, was wrong, a fool's belief, etc., but the ugly beast keeps raising its head. It just won't stay dead, and you surely wouldn't expect trained scientists to keep breathing life back into the beast, but they keep trying!

Pasteur and others demonstrated that each organism requires parents to produce the young. It has been universally held, since Pasteur, that life always arises from life of the same kind. This is the Law of Biogenesis. The evidence points to this fact; however, that conclusion brings men back to God, and pompous scientists (most of them) have trouble accepting the reality of God.

Darwin wrote of the first living organism being present in "some warm little pond."[2] But Darwin never told us where he got his warm, little pond. In his second edition of **Origin**, Darwin did admit that the first form(s) of life may have been produced by a primary cause. (That's God!)

Darwin's last lines were, "There is grandeur in this view of life, with its several powers, having been originally breathed by the Creator into a few forms or into one."[3] But he eventually decided that the subject was "beyond the scope of man's intellect....The mystery of the beginning of all things is insoluble by us; and I for one must be content to remain an Agnostic."[4] ("Agnostic" is a term coined by his disciple, T. H. Huxley.)

## NO EVIDENCE

But Yockey, a Ph.D. from Berkeley, has decided that, after looking at the information content of the least complex life, spontaneous generation did not happen. He wrote that, "...evolution to higher forms could not get started. Geological evidence for the 'warm little pond' is missing....Therefore a belief that proteins basic for life as we know it appeared spontaneously in the primitive milieu on earth is based on faith."[5]

Nobel Prize winner, Ilya Prigogine "joined the choir" when he wrote, "The idea of spontaneous generation of life in its present form is therefore highly improbable even on the scale of the billions of years during which prebiotic evolution occurred."[6]

According to Monod, another Nobel Prize winner and biochemist at the University of Paris, the possibility of life arising spontaneously "...was virtually zero."[7] Yockey, Prigogine, and Monod nevertheless persist in believing in spontaneous, evolutionary origin of life.

Dr. Robert Jastrow is a well-known astronomer, geologist and physicist who admits to being, like Darwin, an agnostic, but his perspective on the origin of life is very interesting. He suggests that it is possible that life began on this Earth by a miracle. He believes that life started by God (and outside scientific understanding) or it started spontaneously from chemical reactions.

Jastrow admits that the first possibility is a matter of faith, consequently not subject to the laws of science; however, he says belief in spontaneous generation is **also** a matter of faith that cannot be supported by scientific inquiry! Jastrow asks, "What concrete evidence supports that remarkable theory of the origin of life? There is none," he concludes.[8]

In 1984, three scientists with doctorates in chemistry, geo-chemistry, and materials science wrote **The Mystery of Life's Origins: Reassessing Current Theories,** which is still making waves and swamping the science departments at major universities. The book throws body blows to the theory that life started by chance. Dean Kenyon, a professor of biology at San Francisco State University wrote the foreword. Kenyon admits that many scientists refuse to study problems with evolution because it "would open the door to the possibility (or the necessity) of supernatural origins of life."[9]

Even the **Yale Journal of Biology and Medicine** and the **Journal of College Science Teaching** have been generous with their praise for **Mystery**: "The volume as a whole is devastating to the relaxed acceptance of current theories of abiogenesis [chemical evolution]."[10]

The theory of chemical evolution maintains that the first cell evolved from a "chemical soup" that was rich in amino acids (and additional organic substances). One of the best known critics of this position is the British astronomer Sir Fred Hoyle who claims that believing the first cell developed by chance is like believing a tornado, sweeping through a junkyard filled with airplane parts, could result in a Boeing 747![11]

Hoyle is an astronomer who deals with probabilities and numbers. He wrote of the spontaneous generation of life appearing by natural causes: "[T]he combinatorial arrangement of not even one among the many thousands of biopolymers on which life depends could have been arrived at by natural processes here on the Earth. Astronomers will have a little difficulty at understanding this because they will be assured by biologists that it is not so, the biologists having been assured in their turn by others that it is not so. The 'others' are a group of persons who believe, quite openly, in mathematical miracles."[12] But then, secular miracles are acceptable to many evolutionists.

We are told that the right chemicals finally collected in some dark swamp (or in the ocean), and after billions of years, life happened through spontaneous generation. But, chemicals don't work toward evolution, only against it! D. E. Hull admitted this when he said, "The conclusion from these arguments presents the most serious obstacle, if indeed it is not fatal, to the theory of spontaneous generation. First, thermodynamic calculations predict vanishingly small concentrations of even the simplest organic compounds. Secondly, the reactions that are invoked to synthesize such compounds are seen to be much more effective in decomposing them."[13]

R. Shapiro, Professor of chemistry and an evolutionist at New York University, seems to be a little embarrassed with all the speculation about life arising spontaneously in a "warm little pond" somewhere, somehow. He wrote, "We have reached a situation where a theory has been accepted as fact by some, and possible contrary evidence is shunted aside. This condition, of course, again describes mythology rather than science."[14] Mythology, magic or miracles!

Brooks and Shaw tell us that there is no proof for the "soup" (in the warm, little pond) anywhere on Earth.[15]

## NO SIMPLE CELL

We used to talk of the "simple" cell, but the electron microscope has changed our terminology. No cell is "simple." It is amazingly complex with the various parts interdependent on the others and each part interacting with many other parts. The cell is an incredible design, requiring a Designer! In each cell, all the parts are separate units, working in unison; however, no part is alive in itself. Life does not exist apart from the complete cell. So how could it have evolved?

Each cell is so complex and information laden that it, alone, is a major support for creation. Dr. Carl Sagan unwittingly supported that fact when he said that the **least** complex cell has information, "comparable to about a hundred million pages of the **Encyclopaedia Britannica**."[16]

Haskins, in an article in **American Scientist**, grudgingly admitted that fact. He wrote, "Did the code and the means of translating it appear simultaneously in evolution? It seems almost incredible that any such coincidences could have occurred, given the extraordinary complexities of both sides and the requirement that they be coordinated accurately for survival. By a pre-Darwinian (or a skeptic of evolution after Darwin) this puzzle surely would have been interpreted as the most powerful sort of evidence for special creation."[17]

In 1953, Dr. Stanley Miller caused a sensation when he showed that amino acids can be produced in the laboratory. This gave the humanist fanatics **hope**! Life from non-life! But alas, it was not to be. Miller only produced a very few of the humble amino acids under **very controlled** circumstances. Even if he had produced the "stuff of life," it would only have proved that it took intelligence to accomplish it--along with generous federal grants!

This rather impressive achievement by Miller in the realm of biochemistry had no real relevance to the origin of life on Earth, and astronomer Fred Hoyle supported that conclusion when he observed, "...the next step--the coming together of subunits into larger organized molecules with the capacity to reproduce themselves--has not occurred in the laboratory flask....As for the appearance of complex biochemicals in the primordial soup, there is an enormous gap in the evidence, one that seems unlikely ever to be bridged."[18]

Much has been made of genetic engineering in recent years, but it is not the creation of life in a test tube. There is a big difference between creating new life and tinkering with existing life!

Dr. Carl Sagan admitted this when he said, "The production of organic molecules necessary for the origin of life is not at all the same as the origin of life."[19]

Dr. Sol Spiegelman, of the University of Illinois, was

successful in synthesizing biologically active RNA, but he began his process with existing viral RNA and a specific enzyme from a virus. When he was asked if he had created life in a test tube, he replied, "Only God can create life."[20]

## GOD DID OR DID NOT

Sir Ernst Chain received the Nobel Prize for his work with penicillin and said, "...the notion that man is about to compete with God is absurd..."[21]

Sir Fred Hoyle, a greatly respected scientist, has been an agnostic, and his sometimes collaborator, Chandra Wickramasinghe, Professor of Applied Mathematics and Astronomy at University College, Cardiff is a well-known atheist. But times are changing. Newspapers in Australia reported in a feature article that both are "changed men. They are both believers."[22]

The article continued, "What convinced both men were calculations they each did independently into the mathematical chances of life starting spontaneously. When each had finished, they looked at the answer almost in disbelief. Each found that the odds against the spark of life igniting accidentally on earth were staggering--in mathematical jargon '10 to the power of 40,000.' "[23]

That translates to "impossible" and convinced two of the most respected scientists in the world that in the beginning there was not "some warm little pond" out of which man crawled.

Wickramasinghe is still uncomfortable with his new position, stating, "I am quite uncomfortable in the situation, the state of mind I now find myself in. But there is no logical way out of it...We used to have open minds; now we realise that the only logical answer to life is creation--and not accidental random shuffling."[24]

Well, have these two men trusted Christ as Savior, and joined a Bible preaching church? Afraid not. They seem to still believe in evolution, but not in Darwin. They believe that life, not existing by accident, arrived in this world from space. They aren't too sure what God is.

Hoyle and Wickramasinghe and other scientists, realizing that spontaneous generation is nonsense, have "reached for the stars" for an answer. Maybe life came to Earth from space! (Of course, that still doesn't solve the problem of its **origin**.) Evolutionist G. G. Simpson said that it is extremely improbable, almost to the point of impossibility that any form of life has ever traveled by natural means from one planetary system to another.[25]

It is suggested that living spores came floating to Earth from space as a result of radiation pressure; however, ultraviolet radiation would have destroyed the spores during their long journey through space.

Others tell us that life came to Earth on meteorites, but such life could not have survived the heat generated when the meteorites struck Earth's atmosphere. Next, these desperate scientists may suggest that beings from space brought the first "Adam and Eve" to Earth in flying saucers to begin our species. Or, maybe it's back to life arising accidently from non-life.

Hoyle and Wickramasinghe wrote of the impossibility of life coming into existence accidently: "No matter how large the environment one considers, life cannot have had a random beginning. Troops of monkeys thundering away at random on typewriters could not produce the works of Shakespeare, for the practical reason that the whole observable universe is not large enough to contain the necessary monkey hordes, the necessary typewriters, and certainly the waste paper baskets required for the deposition of wrong attempts. The same is true for living material."[26] There goes the primordial soup!

Some major scientists are willing to admit that they have no idea how life first started. Luther D. Sunderland said in a taped interview with Dr. David Raup, of the Field Museum of Natural History in Chicago, that, "Neither Dr. Patterson [of the British Museum of Natural History] nor Dr. Eldredge [of the American Museum of Natural History] could give me any explanation of the origination of the first cell." Dr. Raup replied, "I can't either."[27]

Either God created everything or He did not. This either/or position is proposed by Nobel Prize winner in science, Dr. George Wald in the most incredible statement I have ever read by an intelligent person. Wald wrote: "When it comes to the origin of life on this earth, there are only two possibilities: creation or spontaneous generation (evolution). There is no third way. Spontaneous generation was disproved 100 years ago, but that leads us only to one other conclusion: that of supernatural creation. We cannot accept that on philosophical grounds (personal reasons); therefore, **we choose to believe the impossible**: that life arose spontaneous by chance."[28] (Emphasis added.) And **this** is a Nobel Prize winner?

**That** is science? Wald must believe the impossible because he dare not accept the truth! Why? Because it is contrary to his religion, the religion of evolution.

Wald, poor man, suggested that he would be extracted from his paradoxical position by time. He wrote, "Time is in fact the hero of the plot. The time with which we have to deal is of the order of two billion years. What we regard as impossible on the basis of human experience is meaningless here. Given so much time, the 'impossible' becomes possible, the possible probable, and the probable virtually certain. One has only to wait; time itself performs the miracles."[29] No, I'm afraid it takes much more than time.

Wald made another interesting comment on this subject when he said that scientists of a century ago believed in spontaneous generation as a philosophical necessity. He whined that because of the philosophical poverty of our time that necessity is no longer appreciated. He concluded, "Most modern biologists, having reviewed with satisfaction the downfall of the spontaneous generation hypothesis, yet unwilling to accept the alternative belief in special creation, are left with nothing."[30]

Well, they are left with a good salary for teaching falsehoods to impressionable young people. They should be driving a truck where they can't do too much harm!

I'll wind up this chapter with another statement by George Wald, and remember that he is a Nobel Prize winner! (I really don't mean to "pick on" George, but he has said and written so many **dumb** things, I can't help it.) Surely he blushed as he wrote: "One has only to contemplate the magnitude of this task to concede that the spontaneous generation of a living organism is **impossible**. Yet here we are--as a result, I believe, of spontaneous generation."[31] (Emphasis added.) My, my, he's a **real** believer!

The world's leading scientists have no reasonable

238 Evolution: Fact, Fraud, or Faith?

explanation about where the sun, moon and Earth came from. Some suggest that the Earth and the planets were torn from the sun, but no one tells us where the sun came from. Then we are to believe that the right molecules somehow came together, and after a few billion years, life came into existence! But evolutionist George Marsden was correct when he wrote, "Evolution...strains popular common sense. It is simply difficult to believe that the amazing order of life on Earth arose spontaneously out of the original disorder of the universe."[32]

But we are expected to believe, contrary to science, the Scripture and common sense, that life came from non-living matter; however, no sensible, sane person believes such nonsense as spontaneous generation.

# CHAPTER SEVENTEEN

# DARWIN AND NATURAL SELECTION

After life arose "spontaneously," the mechanism that produced man out of molecules was natural selection, according to Darwin and many evolutionists. Evolutionists are quick to say that everyone agrees on the "fact" of evolution, it's only the mechanism that is in disagreement. They think that **saying** it makes it so! But that's like saying that an accused man is not on trial to ascertain the **fact** of his guilt, but to discover the "mechanism" of his alleged crime![1]

Natural selection is the process whereby the creatures that are best adapted to local environments produce and leave more surviving offspring which spread their favored traits through populations.[2]

We are told that many more individuals are produced within each species than can live because of disease, limited food supply, predators, etc., so there is a constant struggle for existence among the various members. Darwin told us that those who possess the most advantageous characteristics would be more competent to survive and produce more offspring. It was a constant struggle for life with the strong exterminating the weak. It was the "survival of the fittest."

Of course, the fastest, most agile, and strongest creatures will survive longer than the sick, the weak and the crippled. But while natural selection may explain the survival of the fittest in the plant and animal kingdom, it does not explain the **arrival** of the fittest!

Species have been disappearing for thousands of years, and evolutionists don't have the answer. Dr. David Raup spoke to this issue: "For every species now in existence, roughly ninety-nine have become extinct. The question of why they have become extinct is of enormous importance to evolutionists. It has been studied by many men, but a convincing answer has not been found."[3]

Raup continues, "The discussion of survival of the fittest showed that the phrase led to circular reasoning; you survive because you are fit and you are fit because you survive. Discussion of extinction is beset by a similar danger. It is all too easy to say that a species becomes extinct because it fails to adapt, while establishing its failure to adapt only by its becoming extinct: in other words, you die because you are unfit and you are unfit because you die."[4] And around and around we go....! Stop, please! I'm getting dizzy!

There is no doubt that the evolutionist is standing uncomfortably and dangerously on a very shaky platform with the noose of a circular argument around his skinny neck.

Some evolutionists are not embarrassed by this circular reasoning, (or tautology) as J. B. S. Haldane wrote: "...the phrase, 'survival of the fittest,' is something of a tautology. So are most mathematical theorems. There is no harm in saying the same truth in two different ways."[5] But of course, he is **not** saying the truth in two different ways, but saying the same truth twice without explaining anything. My, my, their thinking boggles the mind! Evolutionists are experts in double talk.

## BREEDING

Darwin had read the essays of Edward Blyth on natural selection between 1842 and 1844 and corresponded with him after Blyth went to Calcutta. Blyth, a creationist, did not teach evolution but that species remain constant. He also taught that nature weeded out the weak and sickly to maintain a healthy stock of creatures.

Darwin was a pigeon breeder, and from that work concluded that natural selection was the secret to evolution, calling it, "my deity."[6] Of course, natural selection is true as it relates to plants and domesticated animals because a breeder selects a particular breed he wishes to cross with another **with an end in mind**. He then protects that plant or creature and continues to experiment until his goal is achieved. Of course, in nature, none of that is feasible.

E. Russell, in his, **The Diversity of Animals**, said that it is unfortunate that Darwin used the term, "natural selection," because it has given rise to much confusion. He said, Darwin did so "because he arrived at his theory through studying the effects of selection as practiced by man in the breeding of domesticated animals and cultivated plants. Here the use of the word is entirely legitimate. But the **action of man in selective breeding is not analogous to the action of 'natural selection,'** but almost its **direct opposite**....Man has an aim or an end in view; 'natural selection' can have none."[7] (Emphasis added.)

Zoologists Hughes and Lambert aptly wrote, "We should remember that natural selection is not necessarily real...just because biologists and mathematicians have been talking about it for over 100 years."[8] And, of course, talking about it never will make it true.

Darwin referred to the various breeds of pigeons that had been produced through artificial selection from the wild, ordinary rock pigeon that city dwellers see every day. More fancy pigeons can be bred from that rock pigeon, but the breeding will result back to the rock pigeon. There are limitations as to what man can selectively do. Those limits were set in the book of Genesis by God. Natural selection proves creation, not evolution! Elementary, Watson.

Luther Burbank was the greatest breeder of this century, and he supported the fact of limits within a species. He said, "I know from my experience that I can develop a plum half an inch long or one two and a half inches long, with every possible length in between, but I am willing to admit that it is hopeless to try to get a plum the size of a small pea, or one as big as a grape-fruit....In short, there are limits to the developments possible, and these limits follow a law....there is undoubtedly a pull toward the mean which keeps all living things within more or less fixed limitations...."[9]

Would the continued breeding of mice produce a mouse as large as a moose? Of course not, because they will only grow so large. When mice reach the limit, set by God, they will get no larger no matter how much breeding takes place. All qualified biologists recognize the limitation of breeding but as William Fix suggests, "...they do not like to discuss it much when grinding the evolutionary ax."[10]

Deevey's comments on breeding are significant. He affirmed, "Some remarkable things have been done by crossbreeding and selection inside the species barrier, or within a larger circle of closely related species, such as the wheats. But wheat is still wheat, and not, for instance, grapefruit; and we can no more grow wings on pigs than hens can make cylindrical eggs."[11] But they keep trying to grow wings on pigs!

## PEPPERED MOTHS

Evolutionists talk endlessly about the peppered moths as proof of Darwin's theory, and they get glassy-eyed and salivate when they tell the story. It is supposed to be "present day evolution" at work, but here's "the rest of the story."

Every evolutionary textbook deals with the peppered moths of England as proof of evolution by natural selection. It is here they wax eloquent, but it is simply a big yawn! It is "much 'ado about nothing." In the mid-1800s in England they had light and dark peppered moths (and they still do). The light moths landed on light trees and were difficult to see, so the birds, not seeing the light moths on the light trees, **could** see the dark moths on the light trees. The birds did what they are expected to do. They ate the dark moths.

The dark moths were nearing extinction, then came the Industrial Revolution with hundreds of factories bellowing black smoke from their stacks. Soon, the trees became dark with soot making it difficult for the birds to see the dark moths on the dark trees. Ah, but the light moths, landing on the dark bark could be easily seen, so the hungry birds ate them instead of the dark moths they could not see. Now the light moths were almost extinct. That is supposed to be the "evolution going on today" that all the textbooks boast so much about. It doesn't prove evolution; it only proves that birds eat what they can see, and don't eat what they can't see! But we knew that, didn't we?

Pierre Grasse, the most respected scientist in France said that the peppered moths "either have nothing to do with evolution or are insignificant."[12] But you wouldn't know that by listening to the evolutionists bleat on and on about the peppered moths of England. Of course, they are desperate people who will grab onto anything to give themselves some semblance of credibility.

Concerning the peppered moths, Dr. L. H. Matthews admits, "they do not show evolution in progress, for, however the populations may alter in their content of light, intermediate or dark forms, all the moths remain from the beginning to end **Biston betularia**."[13] In other words, they all remain the same species of moths.

With those facts in mind, read what the **International Wildlife Encyclopedia** wrote about the peppered moths. They are "...the most striking evolutionary change ever to be witnessed by man."[14] This is their **best** case! Wow, they **are** in deep trouble!

No, natural selection does not prove evolution. As a major evolutionist, Professor Richard Goldschmidt, admitted, "Darwin's theory of natural selection has never had any proof, yet it has been universally accepted."[15]

## MUTATIONS

It was about the turn of the century that Hugo DeVries suggested the mutation theory to explain how genetic differences came about so that natural selection could choose the strong and eliminate the weak. Mutations are sudden hereditary changes that occur in the genes or chromosomes of individuals, and those changes are inherited by their offspring. Mutations represent a so-called "mistake" in the transmission of genetic information from parent to progeny by way of the genetic code.

A mutation is an unexpected and random change in a cell, produced by the penetration of the cell by radiation, mutagenic chemical or other disorganizing agent. Mutations produce changes, but not improvements.[16]

Some scientists believe that **most** mutations are harmful while others believe they are **all** harmful or neutral. Evolutionists must claim that some mutations are helpful because basically, all evolution rests upon that premise. Grasse, of France, disputed the value of mutations when he wrote, "...No matter how numerous they may be, mutations do not produce any kind of evolution."[17]

Grasse was supported by Professor N. H. Nilsson of Lund University who said, "There is no single instance where it can be maintained that any of the mutants studied has a higher vitality than the mother species."[18] Nilsson added, "It is **therefore, absolutely impossible to build a current evolution on mutations or on recombinations.**"[19] (Emphasis his.)

Michael Pitman, former chemistry professor at Cambridge, confessed, "Neither observation nor controlled experiment has shown natural selection manipulating mutations so as to produce a new gene, hormone, enzyme system or organ."[20]

Even Darwin knew that there was no proof of beneficial changes: "When we descend to details, we can prove that no species has changed (i.e., we cannot prove that a single species has changed); nor can we prove that the supposed changes are beneficial, which is the groundwork of the theory."[21] What an admission!

Not only are mutations always harmful, but they produce changes in **present** characters, never producing new characters. Mutations are the catalyst for imperfection, not improvement, yet evolutionists scream that they are the explanation for all the varieties we see in the animal and plant kingdoms. They teach that the many changes in combination with the pressure of the environment over billions of years have produced the differences between one-celled amoeba and complicated humans; however, mutations never create: they corrupt.

In fact, mutations do not prove evolution, rather, they **disprove** it! Mutations reinforce and support the Second Law of Thermodynamics which teaches that the natural tendency of all change is to produce a greater degree of disorder.[22] That is the result of all mutations: disorder, defects, disease, deformity, and death.

That statement is supported by J. J. Fried in **The Mystery of Heredity**: "We have to face one particular fact, one so peculiar that in the opinion of some people it **makes nonsense** of the whole theory of evolution: Although the biological theory calls for incorporation of beneficial variants in the living populations, a vast majority of the mutants observed in any organism are detrimental to welfare. Some are lethal, causing incurable diseases or fatal deaths; others are sub-lethal killing off or incapacitating most of the carriers but allowing some to escape; still others are sub-vitaol, damaging health, resistance or vigor in a variety of ways."[23] (Emphasis added.)

The mutation theory is without any scientific

foundation. In fact, Ernst Mayr, a famous evolutionist, believed mutations were the answer to how evolution allegedly took place, but his own experiments with fruit flies proved the opposite. Mayr tried to increase and decrease the number of bristles on fruit flies that normally have 36, but he discovered that the flies died if they had more than 56 or fewer than 25. Repeated experiments of 30 generations proved the fact of limitations, and when the fruit flies were left alone, they were back to 36 bristles in five years![24] Of course change takes place, but it is always limited change and **never for good**.

No fruit fly, peppered moth, nor any other creatures have formed a new species through mutations and natural selection, and more and more top scientists are supporting that position. The co-holder of the 1945 Nobel Prize for developing penicillin, Sir Ernst Chain, called natural selection and chance mutations an "hypothesis based on no evidence and irreconcilable with the facts." He continued, "These classical evolutionary theories are a gross over-simplification of an immensely complex and intricate mass of facts, and it amazes me that they are swallowed so uncritically and readily, and for such a long time, by so many scientists without a murmur of protest."[25]

Well, the murmur continues. The 1971 winner of the Nobel Prize in science, Dr. Dennis Gabor said: "I just cannot believe that everything developed by random mutations..."[26] Neither can I, and many honest scientists agree.

In 1967, a group of prestigious biologists and mathematicians convened at the Wistar Institute to consider "Mathematical Challenges to the Neo-Darwinian Interpretation of Evolution." Everyone present was an evolutionist, and they agreed on discussing one question: Is it possible for mutations to serve as the basis with natural selection as a mechanism for evolutionary change from simple to complex organism? The answer of the mathematicians was no! It was not mathematically possible![27]

Dr. Etheridge, world-famous paleontologist of the British Museum said, "Nine-tenths of the talk of evolutionists is sheer nonsense, not founded on observation and wholly unsupported by facts. This museum is full of proofs of the utter falsity of their views. In all this great museum, there is not a particle of evidence of the transmutation of species."[28] What an indictment of evolution; however, evolution still hangs on by a single  frayed thread--mutations.

All right, some honest scientists admit that evolution is a farce, others admit it's a fraud, and still others admit it is a faith. It is pure religion.  It is an attack upon God (what folly!) and an attempt to replace the God of the Bible with the guess of Darwin.

The late Sir Julian Huxley, a modern evolutionary leader, admitted Darwin's purpose in natural selection was to remove the idea of God from any discussion of origins. Huxley said, "Darwin pointed out that no supernatural designer was needed; since natural selection could account for any known form of life; there is no room for a supernatural agency in its evolution."[29]

However, the scientific evidence proves otherwise, much to the shock of evolutionists. The observed facts support special creation. Yes, natural selection does take place, but it never creates nor does it upgrade or change to another species.

This fact was admitted by Dr. Colin Patterson, Senior

Paleontologist at the British Museum of Natural History when he said, "No one has ever produced a species by mechanisms of natural selection. No one has gotten near it...."[30]

Mutations kill off excessive numbers of creatures to keep the environment in balance. Yes,  mutations do take place, but they are **always** harmful, and the facts prove that God has placed a limit to the amount of change that can take place.

The principle of limits is taught in Genesis when God commanded that each plant, animal and man would reproduce "after its kind."  This principle is also taught in I Corinthians 15:38, 39: "But God giveth it a body as it hath pleased Him, and to every seed his own body. All flesh is not the same flesh: but there is one kind of flesh of men, another flesh of beasts, another of fishes, and another of birds." And **that** settles **that**.

Because God set the limits as to the amount of change within a species, it is **impossible** for a creature to evolve into a new creature, but evolutionists are **determined** to find the links that never existed! And that, sadly, has led to much embarrassing fraud among evolutionists as we shall see in the following chapter.

# CHAPTER EIGHTEEN

# SCIENTIFIC FRAUDS

"Evolution must be true," says the naive one, "because natural museums are full of ape men, proof that man evolved from lower forms of life." No, those museum models are reconstructions from very few bones. Ape men have been the hope of evolutionists for more than a hundred years, but they always turn out to be **either man or ape**. Not **one** "missing link" has ever been found anywhere at anytime by anyone! In fact, you could take all the alleged "links" and place all their bones in a single coffin! If ancient ape men roamed the Earth, grunting at each other, and dragging their women by the hair into dirty caves, then where are all their remains?

Lord Solly Zuckerman, the famous British anthropologist, and not a creationist, stated that there are no fossil traces of a transformation from an ape-like creature to man.[1] There are none because it didn't happen, and all the wishing by evolutionists won't make it true.

Some evolutionists act as uncomfortable as a dog in hot ashes when we talk of man coming from monkeys (or apes), and some allege that Darwin did not teach that. They are wrong and dishonest! Darwin wrote, "The Simiadae then branched off into two great stems, the New World and Old World monkeys; and from the latter at a remote period, Man, the wonder and the glory of the universe, proceeded."[2] Well, that should settle whether Darwin taught monkey to man evolution!

One of Darwin's most vocal supporters put it even stronger. Dr. G. G. Simpson seemed to be angry when he wrote, "On this subject, by the way, there has been too much pussyfooting. Apologists emphasize that man cannot be a descendant of any living ape--a statement that is obvious to the verge of imbecility--and go on to state or imply that man is not really descended from an ape or monkey at all, but from an earlier common ancestor....Since the terms ape and monkey are defined by popular usage, man's ancestors were apes or monkeys (or successively both). It is pusillanimous if not dishonest for an informed investigator to say otherwise."[3]

When it comes to protecting, promoting and praising Darwinism, Simpson seems to be mean as a junkyard dog! It would be helpful if he had offered some evidence instead of enthusiasm as proof for evolution.

Most evolutionists believe that men and apes evolved from a common ancestor. Some say this occurred as long as 20 million years ago while others say it may have occurred as recently as five million years ago. Now, if that happened, it means a major shift or realignment of our thinking and theology. If man is only a little higher than the apes (rather than a little lower than the angels) then it is not unreasonable for him to live like an ape. He can forget about sin, righteousness, character and judgment to come. He can spend his days grunting at others, dragging women by the hair into dirty "pads" and generally acting like a brute. He is only acting according to his nature.

Creationists believe that man is a direct creation of a

personal God a few thousand years ago. We maintain that the few bones found of "pre-men" were either apes or men or some other specific creature. There is absolutely no doubt that the sketches and models of ape men are frauds or at best, the results of an overactive imagination. (Or, maybe they were smoking those "funny" cigarettes!)

"But how can honest scientists tell us that a particular 'man' was really a forerunner of man without evidence?" Well, it's called, "wishful thinking." Remember that unbelieving scientists must never admit the **possibility** of God, so when they find a few bones they give them a Greek name (more impressive), call a news conference and boast about finding "further proof of man's evolutionary past." And the reporters are too uninformed to ask the penetrating questions necessary to make the scientists reveal the facts.

Tim White, an anthropologist at Berkeley, rightly said, "The problem with a lot of anthropologists is that they want so much to find a hominid that any scrap of bone becomes a hominid bone."[4]

Evolutionists, because of their bias, find what they **want** to find. In his book, **Origins**, Richard Leakey exposes the Leakey tradition, where "...you look and look again until you find what you know must be there."[5] That is science?

A good example of the non-thinking mind set of most evolutionists is a confession made by Dr. David Pilbeam of Yale University, one of the top physical anthropologists in America. Dr. Pilbeam wrote, "In the course of rethinking my ideas about human evolution, I have changed somewhat as a scientist. I am aware of the prevalence of implicit assumptions and try harder to dig them out of my own thinking. I am also aware that there are many assumptions I will get at only later, when today's thoughts turn into yesterday's misconceptions. I know that at least in paleoanthropology, data are still so sparse that theory heavily influences interpretations. Theories have, in the past, clearly reflected our current ideologies instead of the actual data."[6]

What a confession! Note his last sentence: "Theories have, in the past, clearly reflected our current ideologies instead of the actual data." There are many examples of this fact.

This is a good place to remind you of Professor Ernst Haeckel who faked evidence to support his religion of evolution. You may remember that he admitted that hundreds of the best biologists did the same thing! That's fraud on the part of some, but not all scientists.

## LIFE FROM SPACE

A fragment of meteorite was reportedly found in southwestern France in 1864. The claim was made that it contained evidence of life from space, but it was not looked at closely for over 100 years! (Can you believe scientists would not pursue **proof** of life from outer space if they had it?) The fact is they **needed** some proof to support Darwin's recently published book.

The great scientist, Louis Pasteur had just delivered a strong "defense of divine creation as the only possible source of life." And the meteorite fragment "proved" Pasteur wrong. But it was a fraud! A French microbiologist found that it had been altered by glue, seeds, coal, gravel and tissues to impress the gullible. While it was fraud, I should point out that a French scientist revealed the fraud.

## ARCHAEOPTERYX

Evolutionists have told us that **Archaeopteryx** is a missing link between reptiles and birds, showing a gradual change from one to the other. Of course, they would need to produce thousands of fossils to show the change from one species to another.

Evolutionists put a great amount of faith in **Archaeopteryx** only to be disappointed again. **Archaeopteryx**, according to six British scientists, is a fake and fraud! That's right, Arch has turned out to be a dirty bird. Or to put it another way, Arch fouled his nest and theirs! The report said: "The controversy started when six scientists, including Sir Fred Hoyle, a British astronomer, asserted in a scholarly paper in March that the feather impressions of the museum's specimen had been fabricated in the 19th century hoax."[7] Note that this fraud took place only two years after Darwin published his book, so it tended to provide much needed "proof."

The report continued, "One of the authors, Dr. Chandra Wickromasinghe, an astrophysicist, has been quoted in a British newspaper as saying the purported hoax was carried out by someone who made a paste of crushed limestone from the same period, smeared it round a genuine reptile fossil and then imprinted the feathers."[8]

It is interesting that Hoyle and Wickramsinghe wrote a book that documented this fraud in detail, and both are highly respected scientists. We creationists are appreciative of their contribution to our cause, but then it was a contribution to science in general. Most evolutionists will not look at the truth if that truth shatters some of their pet theories.

## JAVA MAN AND NEANDERTHAL MAN

One good example of actual data being rejected for ideology is the Java Man, discovered by Eugene Dubois in 1890. They dug out a small part of an upper skull, a tooth socket, three teeth and the left thigh bone, and those few bones were not found together! They were 70 feet apart! Of course, there was no way to know that they had made up the same "person." They were also found over a year's time, but that fact was not revealed until much later.

You wouldn't know it from the textbooks, but the experts disagreed over Java. The German zoologists believed "he" was an **ape**, the English thought "he" was **human** and the French thought "he" was an **ape man**! From the meager remains, the scientists made some amazing deductions. One evolutionist said that Java had a tail and another said that his jaw projected like a snout! That's called "creative reconstruction," or more precisely, it is quackery!

But quackery abounds in evolutionary circles. Now comes Haeckel who had a life-size model of Java constructed that was exhibited in museums across Europe. A detailed drawing was made showing the hair on his head, and any intelligent person seeing Java in the papers had no thought that they were looking at a "portrait of a thigh bone, a few teeth, and fragment of a cranium."[9]

But the **big** secret that Dubois kept from everyone is that he had found two human heads at the same level at which Java was found! Of course, that would have eliminated Java as ape man since men were already living!

After all, you can't live before your parents do, except in the fairy tale world of evolution.

Eugene Dubois, the discoverer of Java Man, finally admitted just before his death that Java was not an ape man at all. The skull was really the remains of a large gibbon (tail-less ape)! After he blew the whistle on this fraud, some evolutionists continued to promote Java in the school textbooks as a "person" who had actually lived. Other evolutionists dismissed Dubois as unreliable because of his admission of fraud. Real people of character, huh?

Why don't the scientists ask some pertinent questions in regard to uncovered bones? (Of course, some scientists did reject Java as ape man, but they were lost in the crowd that applauded the discovery because it supported Darwin when he needed the support.)

How could a handful of bones be used to reconstruct a prehistoric race? How could they be sure that the bones belonged to the same "individual"? How have those **unpetrified** bones survived for the alleged million years?[10]

But evolutionary scientists don't ask the hard questions that would reveal the fact that evolution is resting upon nothing except hot air, imagination, wishful thinking, circular reasoning and fraud! Many of the evolutionists must write part-time for the Saturday morning cartoon shows. After all, making deductions about Java's "snout" and "tail" from the meager evidence discovered is like finding the rear wheel of a 1959 Edsel in a ditch along with the left front headlight, and arriving at the conclusion that the original Edsel had been driven 76,000 miles at an average speed of 41 miles per hour by a little old lady from Pasadena--with fallen arches! Scientists indeed!

"Neanderthal" has become a well-used word in the English language. It is often used to denigrate Christians or conservatives because of our "antiquated" views. The word carries with it the idea of crudeness, ignorance and a lack of sophistication. In other words, a "Neanderthal" was not much of a man; maybe a half man!

Most people made a "big deal" about Neanderthal Man who was first found in Germany in 1856 with a brain capacity **larger** than modern man. Yet, he was depicted as a barrel-chested, hairy, brutish, stooped person, intermediate between ape and man. (Maybe the original big city mugger!) Now we know otherwise. He was not ape man, but a **man** with arthritis and rickets, who buried his dead with ceremony (indicating some religious belief) and was a skilled craftsman.

Anthropologist A. Montagu of Princeton University agreed: "Owing to want of a little knowledge of elementary anatomy, some of these 'authorities' who have engaged in 'reconstruction' of Neanderthal man have represented him with a bull neck, grotesque features, and walking with a stoop....It has also often been asserted that Neanderthal man must have been of low intelligence because he had a low forehead. All these slanders are indefensible."[11]

As early as 1864, the German anatomist Mayer called attention to Neanderthal's physical problems (in addition to being dead!) and in 1872, Rudolf Virchow, father of modern pathology, boldly said that Neanderthal was **not** ancient but a human who suffered from rickets.[12] That was in the late 1800s!

But the evolutionists didn't want to give up Neanderthal, so respected scientist, Marcellin Boule from the National Museum of Natural History in Paris, tried to reconcile the large brain capacity (indicating advanced intelligence) with him being a brutish man. Boule decided to use the pseudoscience of phrenology (using the bumps on the surface of the skull to judge intelligence) and argued that, yes, Neanderthal had a large brain, but it was inferior-- according to phrenology. If bumps on the skull indicate intelligence, most evolutionists are old smoothies!

So where did the statues and pictures, showing a stumbling brute come from if Neanderthal was full man who would not be looked at twice on the street tomorrow morning? (All recent "pictures" depict him as a modern man, although stooped from arthritis.) They were pulled from the minds of artists, after all, an artist can make Neanderthal look like a savage or a scholar depending upon his bias. (E. A. Hooton said that the alleged restorations of ancient types of men have very little, if any, scientific value and are likely only to mislead the public.)[13] No doubt about that.

By 1975, Neanderthal was recognized as fully human. So Neanderthal was not half-man, but full man with a debilitating disease. Down goes another "missing link."

## PILTDOWN MAN

I would be remiss if I did not deal with the famous, now infamous, Piltdown Man. "Pilty," if I can be so irreverent, turned out to be a fraud. Charles Dawson, a Sussex lawyer, had dug up some bones (said to be half a million years old) in 1912 near Piltdown, England that made each evolutionist's heart beat faster than a liberal politician's heart at the annual meeting of the American Civil Liberties Union. The evolutionists, flushed and excited, pointed to their "missing link" for more than 40 years, and boasted about it in all the textbooks as being "proof of evolution." In fact, more than 500 doctoral dissertations were written on Piltdown Man, seeking to establish "him" as one of the most impressive "finds" in the history of science.

What did they find that made them so excited? A jawbone and part of a skull! Then the jawbone was judged to belong to an ape 50 years old, not half a million years old! The skull was also young in that it was found to be only a few thousand years old, not millions! When this became known, the "experts" (anyone fifty miles from home with a briefcase) looked more closely at the "fossils" and discovered fraud. Yes, I know scientists are supposed to be above that, but remember that the human heart is depraved and capable of incredible sin.

In 1953, three honest scientists found that the modern ape's jaw and human skull had been "doctored" to resemble an ape man, and the fake fooled the world's greatest experts! The persons responsible for the hoax (fraud) some allege, were Dawson and Pierre Teilhard de Chardin who had written several books trying to harmonize evolution and Christianity. (Teilhard was a Jesuit priest.) Evidently he decided that Darwin and his disciples needed help to prove their point; however, in "professing themselves to be wise, they became fools" (Romans 1:22).

G. Himmelfarb candidly admitted the problem with Piltdown. "The zeal with which eminent scientists defended it, the facility with which even those who did not welcome it managed to accommodate to it, and the way in which the most respected scientific techniques were soberly and painstakingly applied to it, with the apparent result of confirming both the genuineness of the fossils and the truth of evolution, are at the very least suspicious."[14]

The fraud engulfs even the officials of the British Museum, who, with missionary zeal, kept unfriendly eyes from examining the Piltdown "remains" for 40 years!

This cleverly devised hoax made the scientists look like inept, incompetent and inexpert fools, and Pilty was finally dropped (after pressure from creationists) quicker than a failing television series. Well, the evolutionists have their Dawson and Teilhard, and we Christians have our Jimmy Swaggart and Jimmy Bakker.

Dr. W. R. Thompson, an evolutionist, commented on the Piltdown and Java men in his introduction to the 1956 edition of Darwin's **Origin of Species**: "The success of Darwinism was accomplished by a decline in scientific integrity....A striking example...is the alteration of the Piltdown skull so that it could be used as evidence for the descent of man from the apes; but even before this a similar instance of tinkering with evidence was finally revealed by the discover of **Pithecanthropus**, who admitted, many years after his sensational report, that he had found in the same deposition bones that are definitely human. Though these facts are now well known, a work published in 1943 still accepts the diagnosis of **Pithecanthropus** given by Dubois, as a creature with a femur of human form permitting an erect posture."[15]

I accept Thompson's candor, but if a Christian creationist had been involved in such fraud, it would not be called, "tinkering with evidence." It would be "fraud," "dishonesty," "hoax," etc., and he would be forced to resign whatever position he might have. But with evolutionists, it is "tinkering."

## NEBRASKA MAN

Nebraska Man, found in Nebraska in 1922 by Harold Cook, was a major discovery, but what did they find? A single tooth! I kid you not. Professor Henry Osborn, of the Natural History Museum in New York, in a speech to the American Philosophical Society of Philadelphia on April 28, 1927 opined that Nebraska Man was the **earliest** fossil man ever found in North America. Consequently, it belonged to an ape man according to the "experts."

While it was only a tooth, imaginative scientists put the tooth inside a skull and the skull on a skeleton, then dubbed him, **Hesperopithecus haroldcookii**. (Doesn't that sound impressive?)

One professor said the tooth was worth a million dollars! Nebraska Man's "picture" appeared in the **Illustrated London News** on June 24, 1922, with "him" sitting around his cave **with his wife**! Wasn't that some deduction? They produced the likeness of Nebraska Man **and** his wife from a single tooth! Isn't it amazing what scientists can do with the cooperation of the media?

Well, the pictures, the press, and the professors all impressed people, even uninformed Christians. Some of them fled to their churches, clutched their Bibles, and whimpered unconvincingly about how "the Bible is still true--in the original languages, but maybe we need to look again at Genesis one and two. Maybe there is something to the gap theory, after all."

But when they unearthed "his" skeleton, Nebraska

Man turned out to be an extinct wild pig! And the newspapers did **not** give that fact much coverage! Wonder why? Dr. Duane Gish wrote of this fraud: "I believe this is a case in which a scientist made a man out of a pig and the pig made a monkey out of the scientist."[16]  I **like** that.

In the famous Scopes Monkey Trial of 1925, in Dayton, Tennessee, Nebraska Man was to speak loudly for the defense. Most people are only aware of that trial from a Hollywood production, but the film was riddled with fantasy, falsehood and fiction. (As usual, Hollywood monkeyed around with the truth.) The idea of a legal confrontation was hatched in the inner sanctum of the New York office of the American Civil Liberties Union! The ACLU advertised in the **Chattanooga Times** for a teacher to test the state law that prohibited the teaching of evolution.

John Scopes was presented by Hollywood, as a fearful, surprised defendant, but the facts are different. He was a basketball coach and substitute science teacher, and he wasn't even sure he had taught evolution! At a meeting in a Dayton drugstore, Scopes agreed to be the defendant, and he tutored two boys in evolution while sitting in the back seat of a taxi in front of the drugstore  so he could "tell the truth" at the trial.

The ACLU and members of the American Association for the Advancement of Science hired the famous Chicago lawyer, Clarence Darrow to defend Scopes and the prosecution was assisted by three times Democrat presidential candidate, William Jennings Bryan who was an outspoken foe of evolution and  an active Christian.

Darrow waved the London newspaper, with the "picture" of Nebraska Man,  and ridiculed Bryan and

Christians everywhere. Hollywood made Bryan look like a fool and crackpot when the fool and crackpot was Darrow as the trial transcript shows--especially in light of what we know about ape men in general and Nebraska Man in particular.

The **Omaha World-Herald** humorously, although properly, chastised scientists in their February 24, 1928 edition: "**Hesperopithecus**, alias, the Million-dollar Tooth, has taken another tumble, and, oh, what a bump it got! Instead of being the sole surviving remnant of a noble Nebraska Neanderthal, it turns out to be a grinder from the prehistoric jaw of one of his pigs. Instead of being worth a million, it is worth, judging from the price of pork on the hoof today, perhaps all of 30 cents."[17]

Now some may think they were unkind, but I think it was apropos. Scientists have been "getting away with murder," and it's time we call their hand. The Omaha editor did just that, much to the delight of creationists.

The Omaha paper continued, "It isn't so much the loss of this tooth's fame which we mourn as it is the loss of our faith in the infallibility of science. If there is so little difference between a pig's tooth and a man's that the one may be mistaken for the other for eight years, aren't there infinite chances for error in the identification of the fossilized fragments out of which such an amazingly strange prehistoric fauna has been recreated for us?"[18] (No one expects scientists to be "infallible"; only honest.)

No one talks about Nebraska Man anymore! "He" has been buried along with his tooth. No doubt most people think that evolutionists should learn from their mistakes and frauds, but they keep showing up.

## PEKING MAN

Then there was Peking Man who arrived on the international scene in 1927. Dr. Davidson Black had been involved with Piltdown and in 1926, he was professor of anatomy at the Medical College in Peking, China. Black received an annual grant from the Rockefeller Foundation to excavate at Choukoutien. In charge of the assignment, Black was assisted by Dr. Pei, a Chinese scientist in charge of the field work. Pierre Teilhard de Chardin, of Piltdown fame, was an advisor to Black. (Teilhard had been censured by the Roman Catholic Church for his fanatical evolutionary views.)

In 1927, (while Pilty was "riding high") a molar tooth was discovered, and Black announced another ape man had been unearthed--**Sinanthropus** or Peking Man. In 1929, a skull was found, and Teilhard reported to Europe that the skull "manifestly resembles the great apes closely." However, the skull had a small brain capacity.

Dr. Black made a model (not a cast) of the skull and wrote a long-winded article trying to "sell" Peking Man as an intermediate between Java Man and Neanderthal Man. Black represented the brain as being in the human range in opposition to Teilhard's description. (Black wanted an ape man, and he was going to have an ape man, one way or the other.)

During further excavation, they discovered two huge heaps of ashes with bones of numerous animals. They also found more small skulls which Black claimed were more of his Peking Man. Since only the skulls were found, it was obvious the monkey-like skulls had been carried into the "cave." There were no fossils to indicate how the creature walked or stood, but Teilhard proclaimed that Peking had walked upright, and was two-handed! My, my, what science can do!

The world was excited about Peking Man, a link with the past! The fact that "traces of fire" had been found proved that "he" was just across the line from animals. Here was Peking Man who used crude tools, walked upright, lived in a cave and used fire for cooking. The press and the world fell for it--again.

However, a renowned paleontologist, Professor Abbe Breuil, visited the site in 1931 and returned to France asking questions that raised some doubts about Peking Man. The "traces of fire" turned out to be huge industrial furnaces, and the professor, seeing the small skulls, asked how creatures with such small brains could have worked such an industry. It was discovered that the "traces of fire" were as long as a football field, half as wide as a football field and 30 feet high. The evidence proved that this was a large industry of converting limestone into lime carried on by advanced humans, probably in the construction of the ancient city of Cambalnc, near present day Peking.

Black and Tielhard did not reveal the extent of the ash heaps, calling them small "hearth fires." Peking Man was represented as living in a cave, huddled around his fire, but there was no cave.

The respected Marcillin Boule, from the National Museum of Natural History in Paris was not an anti-evolutionist, but he was a scientist with character, (although he had been "burned" in the Neanderthal caper when he used phrenology to determine intelligence). He visited the site, but believed that he had wasted his time with monkey skulls. He left the site convinced that modern men had worked a large industry. He said the skulls and other animal bones found in the ashes were only the remains of food eaten by human workers who had tossed the bones into the ashes.

Just when you think things can't get worse, they usually do, and they did for Black and Tielhard. In late 1933, three human skulls were unearthed and more bones were discovered later as their house of cards came tumbling down.

Black was found dead in his laboratory in March of 1934, and Dr. Franz Weidenreich replaced him as top honcho, and he outdid Black! He produced his own model of Peking Man using four different skull pieces, then had a sculptress, Lucille Swann, to mold it into a woman's head with a very thick neck. Evolutionists were limited only by their imagination, and produced for the world an intermediate between ape and man! But it was not to be, and scientists, again, looked like fools, fanatics and fakers.

## RECENT DISCOVERIES

Dr. Louis Leakey and his wife discovered the East Africa Man that was supposed to be almost two million years old according to the University of California's potassium-argon dating process. (They later stated that age to be half a million years too great.)

Two films were made touting this "amazing" discovery; however, Dr. Leakey later admitted that his East African Man (**Zinganthropus**) was a variety of **Australo-pithecus**, discovered in 1924 in South Africa. Most evolutionists still claim that **Australopithecus** was on his way to becoming human but this claim has been challenged by evolutionists Lord Solly Zuckerman and Dr. Charles Oxnard.[19]

Dr. Louis Leakey's son, Richard--now an anthropologist (who did not go to college)--admitted that his famous father had found some extant apes. In fact, he admitted that his own findings destroyed all we have been taught about evolution.[20] Richard had found "Skull 1470" that was classified as **Homo**, but it was dated at three million years. That date was rejected, and brought down to two million years. Leakey has settled on 1.8 million years. It was thought the skull was male, but now it is believed to be female. (Maybe evolutionists should be classified as **Homo ignoramus** since they really don't **know** anything about the origin of man.) Leakey does know that money makes the world go around, so he keeps beating on the doors to raise funds to continue his bone hunting.

Professor Enoch, in his **Evolution or Creation**, commented on Leakey's discovery suggesting that it "has not only blasted all the previous speculations of human evolution from the ape or ape-like ancestors, but also now compels us to conclude that man was a separate creation from the beginning, being older than all the missing links and apes cited by evolutionists."[21]

Leakey, realizing this new information would require a total reevaluation of evolution said, "I have nothing to offer in its place."[22]  Hey, Dick, I do. Try Genesis. But then, evolutionists dare not even consider special creation since they would then have to answer to God. They prefer to deceive themselves into believing in the myth of ape men. They keep looking for the missing links between lower forms of life and man, but they seek in vain. About like a blind man in a dark basement looking for a black cat--that isn't there!

Layer of Moon Dust Disproves Ancient Universe

## CHAPTER NINETEEN

# HOW OLD ARE THE WORLD AND THE UNIVERSE?

It is almost universally believed that the universe is billions of years old, yet that is simply an assumption the evolutionists have made to accommodate their outrageous unscientific theories. Few people know that human records only go back about five thousand years. If you go back any further, you must use a great amount of speculation and assumptions. The evolutionist's age for the universe, the Earth and man are pure guesses--and not good guesses at that!

The evolutionists must have time--a long, long time--to paste together their myth of gradual development. It should be stated that we could have an ancient Earth without evolution being a reality, but you can't have evolution without very long periods of time. They almost put creationists in the "flat earth" category when we insist on a young Earth. After all, isn't it obvious that the Earth is old? No, it isn't. What "clocks" are you looking at? When we ask that question, they always point to the fossils and rocks, but, as I have dealt with elsewhere, that "clock" **doesn't work**. What good is an unreliable clock?

Yale geologist, Carl Dunbar, lays the foundation for evolution: "Although the comparative study of living animals and plants may give very convincing circumstantial evidence, fossils provide the only historical, documentary evidence that life has evolved from simpler to more and more complex forms."[1]

But how do we **know** when various animals lived back in time? We can't know when they lived, how they died or under what circumstances they died from any written record. So evolutionists have come up with a **theory** (read: guess) as to what happened. Of course, any sane person would admit that there might be many theories as to what occurred "millions of years ago."

Dunbar says, "Since fossils record life from age to age, they show the course life has taken in its gradual development. [This is an **assumption**, a **belief**, a **hope**, a **theory**, a **postulate**, an **hypothesis**, a **supposition**, etc.] The facts that the oldest rocks bear only extinct types of relatively small and simple kinds of life, and that more and more complex types appear in successive ages, show that there has been a gradual development or unfolding of life on the earth."[2]

He is telling us that as time passed, the record was laid down for us to see today, and it is a consistent system that we can trust. The fossils are always "in place" wherever they are found any place in the world. Since life developed from very simple to complex, you can always find the simple creatures in the oldest strata and the more complex creatures in the recent strata. That is the pitch of evolutionists, but it is deceitful, wrong, untrue, erroneous and inaccurate.

It is the conventional "wisdom," but it is not factual. Dunbar doesn't tell you that it isn't always that neat and precise in the strata. He doesn't admit that the rocks are often out of sequence by "millions" of years, nor does he tell you that creatures appear in the rocks ages before they existed! (To believe **that** can happen requires one to believe in miracles, magic or myths. Take your choice.)

Geologists affirm that if you could pile the thickest sedimentary beds of each geologic age on top of each other in proper sequence, it would reach over 100 miles high! Students are permitted to believe that most or all of the geologic column is available for inspection, but only a mile can be seen, and that is in the Grand Canyon.

Neither is it revealed that below the sedimentary rock layers will be found the "basement complex," and that **any** period of the column can be found on top of that basement layer! Wait a minute, that can't happen; but it does happen frequently. You see, the "clock" is not reliable, so we are getting the wrong "time."

Evolutionists also teach that there are "index" fossils in the strata that tell us the age of the rocks. If a large number of fossilized creatures are found in a stratum, then that gives the age of the rock. But how do we know when the creatures evolved, then died? Well, **by the age of the rock**! Round and round she goes.

Yes, that is circular reasoning, and some intrepid scientists have admitted the fact. R. H. Rastall of Cambridge University said, "It cannot be denied that from a strictly philosophical standpoint geologists are here arguing in a circle. The succession of organisms has been determined by a study of their remains buried in the rocks, and the relative ages of the rocks are determined by the remains of organisms that they contain."[3]

So, while they admit to arguing in a circle, the

scientists tell us that the Earth strata and fossils (that allegedly validate each other!) give us a relative age for the Earth. The rock strata were laid down over billions of years before man evolved; however, if that is true, how did man-made articles get into layers of solid rock? Great question!

The **London Times**, 1851, reported that Hiram de Witt found a perfect iron nail inside a piece of auriferous quartz in California! That can't be according to the evolutionist's "clock."

## CLOCKS NOT WORKING

Another iron nail, six inches long, was found in the sixteenth century by Spanish conquistadors, incrusted in rock deep in a Peruvian mine. The rock was estimated to be tens of thousands of years old, long before man could make iron, according to evolutionists. Their "clock" is still not working!

Then in June of 1851, workmen were blasting near Dorchester, Massachusetts, and they saw a bell-shaped metal vessel blast out of the solid rock! It had inlaid floral designs, and showed a high degree of workmanship.[4] How did the metal vessel get into solid rock if the rock was laid down over millions of years before man ever lived? Their "clock" is way off--by millions of years!

But, surely it takes millions of years for mountains to develop, rivers to cut deep gorges and lakes to be made. No, it doesn't. A massive flood along with earthquakes and volcanoes can easily account for those physical events in a short period of time.

Look at eastern Washington with its channeled scab-lands where more than 15,000 square miles are deeply gouged by steep-walled canyons out of crystalline lava rock. Geologists assumed that these canyons were the result of very slow river erosion over millions of years, but not so! The actual field evidence convinced a United States Geological Survey team that the scablands were formed by the great "Spokane Flood."[5] The time was not millions of years, but in one or two days! The USGS has produced a booklet that admits that fact!

There are numerous examples of the "clock" not working, but one more will suffice. In 1963, a totally new island was formed a few miles south of Iceland, not an unusual occurrence. The next year, Iceland's foremost geophysicist, Sigurdur Thorarinsson, wrote a book about the island. His amazing description follows:

"Only a few months have sufficed for a landscape to be created which is so varied and mature that it is almost beyond belief....Here we see wide sandy beaches and precipitous crags lashed by breakers of the sea. There are gravel banks and lagoons, impressive cliffs resembling the White Cliffs on the English Channel. There are hollows, glens, and soft undulating land. There are fractures, and faulted cliffs, channels and rock debris. There are boulders worn by the surf, some of which are almost round, and further out there is a sandy beach....

"An Icelander who has studied geology and geomorphology at foreign universities is later taught by experience in his own homeland that the time scale he had been trained to attach to geological developments is misleading....What elsewhere may take thousands of years may be accomplished here in one century. All the same he is amazed whenever he goes to Surtsey, because there the same development may take a few weeks or even a few days."[6] Wow! Looks like we creationists aren't the only ones who are critical of the evolutionary "clocks."

So when someone discusses the age of the Earth, you must ascertain what "clock" he is using to tell the time. Remember that a clock is only as good as the assumptions behind it.

## THE DUST CLOCK

Let me give you a good example from my life. After my wife died in 1987, my teenage daughter and I were on our own. Everything was new and tenuous. She was afraid to cook, and I was afraid to eat it. She had to learn to do the laundry, ironing, shopping, cleaning, etc. One day I noticed a heavy layer of dust on top of the refrigerator. Question: how long had it been since she had dusted? I raked my finger across the top of the refrigerator to gauge the amount of dust and estimated how long a period had passed since the last dusting. The amount of dust is a kind of timepiece to measure the time between dusting. Of course, it will not be exact, but I could measure the estimated time between cleaning.

If there is a thin layer of dust, a short time has passed, but if there is a quarter inch of dust, someone has been negligent. But is this **always** true? You can use the layer of dust to make some assumptions, but there are further considerations if you want to make an accurate calculation. If you are remodeling your house, there may be a thick layer of dust that would not reflect careless housekeeping.

Maybe a very odd person **likes** dust (I think I've been

in some of those homes) and actually adds dust to the refrigerator every few days thereby producing an incorrect reading of the "clock." Obviously, a clock is only as good as the assumptions behind it.[7]

Scientists have been telling us for many years that cosmic dust has been settling on the moon for billions of years, and would be an accurate "clock" as to the age of the universe. The moon dust was calculated to be 54 feet thick, and there was great concern for our astronauts when plans were made for our first moon landing.

Some thought the spaceship might land and slowly sink out of sight even a mile into dust! Astronomer R. Lyttleton, consultant to the space program, surmised that there might be a dust layer "several miles deep."[8] Isaac Asimov, atheist science writer, predicted our first manned landing by writing, "I get a picture, therefore, of the first spaceship, picking out a nice level place for landing purposes coming in slowly downward tail-first...and sinking majestically out of sight."[9] Well, it didn't happen. Their "clock" was not working.

You may remember that Bob Hope asked Neil Armstrong on national television what his greatest fear was when he set that first historic foot on the surface of the moon. Armstrong replied that it was the moon dust the scientists had told them to expect. So NASA scientists designed large web-like landing pods to keep the  spaceship on solid ground.

When the spaceship landed, it did not sink into lunar dust, and Neil Armstrong disembarked making his famous statement heard around the world, "That's one small step for man,  one  giant leap for mankind." Then he tried to plant the

American flag into "billions" of years of cosmic dust by hammering the pole, but he hit moon rock instead of dust!

After that incident, the evolutionists started rewriting history because the 54 feet or more of lunar dust was not there. How much did they find? From one-eighth inch to three inches! Sorry about that fellows, but your "clock" is not working--as we creationists have been telling you for years. The moon is young, not old.

Evolutionists began to lie about the moon dust because it showed them wrong once again. They deny they ever took that position on moon dust, but the facts are otherwise, proving they are not only desperate, but dishonest and disreputable scientists.

NASA put millions of extra dollars into that first moon mission because of the moon dust. Obviously they strongly believed in the billions of years of cosmic dust falling on the moon, and made physical changes in the equipment because of it.

The **Rand McNally New Concise Atlas of the Universe** clearly states, "The theory that the Maria were covered with deep layers of soft dust was current until well into the 1960s."[10]

Astronomer Robert Dixon wrote in 1971, "The moon was for many years characterized as having a thick layer of dust covering its surface, into which an object would sink if it landed on the moon."[11] Looks like the evolutionists have been caught with their pants down--again.

## YOUNG EARTH

According to the fairy tale, the geologic column has

been developing over billions of years, and consequently should reveal numerous meteorites that have fallen to Earth from time to time. Most meteors would burn out before reaching the Earth, but many would impact the Earth and would be found in the geologic column. But they aren't there because the rock strata were not laid over billions of years, but in less than a year during and following the Flood of Noah! (Of course, meteorites have been found on the **surface** of the Earth, but never in the strata, as you would expect if evolution had occurred.)

A study dealing with meteoritic evidence showed that there is not one example of meteorites being found in the geologic column.[12] Not one!

While the biblical dates done by Archbishop Ussher are not inspired, they are rather accurate when compared to other "clocks." The fact is, the Earth and the universe are very young--not very ancient--much to the consternation of the evolutionists who **must** have long periods of time to develop their cockamamie story of macroevolution.

John Eddy, a leading high altitude astronomer spoke at a scientific conference at Louisiana State University, and shocked the audience when he admitted that there was not much evidence to disagree with the value of only 6,000 years proposed by Archbishop Ussher. (**GEO Times**, Sept., 1978.)

"But," say some sincere Christians, "everyone **knows** the Earth is billions of years old." How do we know that? Genesis surely does not indicate an old age for the Earth, and neither does science. We have been brainwashed through our education, the media and our associates. We have bought another gold brick sold by thumbsucking liberals.

"Well," says another, "isn't it possible that God could

have created the world over a long period of time?" Of course it is **possible**, and even if one admitted it were also **probable**, that wouldn't make it **true**. God could have done it anyway He wanted, but the evidence and the Bible clearly prove special creation.

According to a group of mathematicians, all of whom were evolutionists, it would have taken not five billion years for man to evolve but **billions** of times longer! So they need even **more** time for their myth.

## DATING METHODS

Various methods (or clocks) have been used in the past to date the Earth such as the rate of influx of salt into the oceans, the rate of recession of waterfalls, the rate of sedimentation, etc. In recent years, more "successful" methods have been devised that are supposed to be more sophisticated and reliable. Yet, Dr. Henry Faul wrote concerning one of those "reliable" dating methods--uranium dating: "...widely diverging ages can be measured on samples from the same spot."[13] From the **same** spot! Very scientific!

It is also true that **young** deposits (168 years old), such as lava beds formed by **modern** volcanoes, have been dated by potassium-argon to be three billion years old![14] It seems the "experts" have made a guessing game out of the dating game.

Stansfield honestly confesses to the unreliability of radiometric dating when he wrote, "Age estimates on a given geological stratum using different methods are often quite different (sometimes by hundreds of millions of years). There is no absolutely reliable long-term radiological 'clock.' "[15]

That position is supported by an article in the **Bulletin of the Geological Society of America**, written by Curt Teichert, "No coherent picture of the history of the earth could be built on the basis of radioactive datings."[16] Well, who ever said that evolutionists were coherent?

Rock samples from twenty-two volcanic rocks in different parts of the Earth are known to have been formed during the last 200 years, yet radio-active dating methods provided ages that ranged from 100 million to 10,000 million years![17] I'd say that clock is a little off!

Some Russian volcanic rocks only a few **thousand** years old were labeled as being from 50 million to 14.6 billion years old! In Haevair, volcanic rocks only 81 years old have been judged to be from 400,000 to more than 3 billion years old!

A major problem in using radioactivity to date rocks is the fact that we can't be sure the decay rates have always been constant. In fact, there is evidence that they may not have remained constant.[18]

Another problem is an assumption of a closed system. A closed system demands that uranium or potassium was in the rocks for "millions" of years without any kind of contamination. However, that requires a massive leap of faith without **any** foundation in science. There can be no doubt that, over the years, leaching, evaporation, etc. took place that throws off the "clock."

A good example of this is seen in the dating of Leakey's Skull 1470 that was initially dated at Cambridge Laboratory using the potassium-argon method. The first date was 221 million years, but it was rejected because it didn't fit the evolutionist's scenario. Further testing produced dates from 2.4 to 2.6 million years. (Many millions of years difference in those dates and the first date!) Then they got another date of 1.8 million years from the University of California, Berkeley. Now, that's more like it. That date fits their fairy tale! Isn't it interesting how they can adjust their "science" to fit their philosophy? (Leakey now sticks to 1.8 million years.)

But further testing produced dates ranging from 290,000 years to 19.5 million years. Eeny, Meeny, Miney, Mo. You take your choice. Now you can understand why some have described the potassium-argon clock as being a clock without hands; without even a face.[19]

The thing that surprises me when an evolutionist spouts his drivel about how they can "prove" the great age of the Earth is that informed people don't fall to the floor, gasping and holding their sides with raucous laughter!

Evidently the dating methods are not as reliable as they have been projected to be, but it gets worse! **Science** magazine reported on the work of Dr. M. S. Cheat and G. M. Anderson, stating that the shells of **living** mollusks were calculated to have been **dead** 2,300 years![20] Well, so much for "sophisticated and reliable" methods of dating the Earth.

**Science** magazine again wrote about, "the failure of radio-carbon technique to yield dates of certain dependability." It said, "Although it was hailed as the answer to the prehistorian's prayer when it was first announced, there has been increasing disillusion with the method because of the chronological uncertainties, in some cases absurdities, that would follow strict adherence to the published Carbon 14 dates."[21] My, my, "uncertainties and absurdities!"

Another absurdity is obvious when wood taken from

**living** trees was dated to be 10,000 years old by the carbon-14 method![22] And mortar from an English castle about 800 years old has been dated 7,370 years![23] But can you believe it gets worse? Seals that were **freshly** killed were given an age of 1,300 years and mummified seals were dated at 4,600 years when their known age was only 30 years old![24]

Robert E. Lee further warned about radiocarbon dating when he admitted, "The troubles of the radiocarbon dating method are undeniably deep and serious. Despite 35 years of technological refinement and better understanding, the underlying **assumptions** have been strongly challenged, and warnings are out that radiocarbon may soon find itself in a crisis situation....It should be no surprise, then, that fully half of the dates are rejected. The wonder is, surely, that the remaining half come to be accepted.

"No matter how 'useful' it is, though the radiocarbon method is still not capable of yielding accurate and reliable results. There are **gross** discrepancies, the chronology is **uneven** and relative, **and the accepted dates are actually selected dates.**"[25] (Emphasis added.)

Note that half the dates are rejected and there are "gross" discrepancies. Question: how could any evolutionist speak with authority regarding dating?

One more devastating fact comes from a meeting of Nobel Prize winners in Uppsala, Sweden in 1969. It was admitted that, "If a C-14 date supports our theories, we put it in the main text. If it does not entirely contradict them, we put it in a footnote. And if it is completely 'out-of-date,' we just drop it."[26] I rest my case on the new "reliable" methods of dating **and** the honesty of many evolutionists who talk endlessly of "billions of years."

## NATURAL CLOCKS

So we can disregard the new, sophisticated methods of long-term dating, and return to the more "natural" methods. But even those are not always reliable unless you have all the facts and they are properly evaluated.

The estimated ages for the Earth (always very old to fit into the evolutionists' fairy tale) are usually based on "clocks" that have been ticking at very slow rates. A good example is the coral growth rates that have been thought to require hundreds of thousands of years, but now it is believed that no coral formation need be over 3,500 years old![27] There are many such proofs of a young Earth, yet blinded evolutionists keep demanding an old Earth.

Similar to coral formations, stalactite and stalagmite formations in caves are said to take long ages to form. But now we know that the evidence is specious, and they can form in only a few years. The cave guide can prate on and on about the ancient formations, but the evidence refutes their canned spiel.

A curtain of stalactites, some five feet in length, grows from the foundation ceiling beneath the Lincoln Memorial in Washington, D.C. The memorial was built in 1923, so those stalactites have been produced in less than a hundred years![28]

Further evidence of non-uniformitarianism relating to stalagmites is found in the Carlsbad (NM) Caverns where a bat was found encased inside a stalagmite! That is impossible if many years are necessary to form stalagmites and stalactites because the bat would have decayed or been eaten by predators in a very short period of time.[29] Looks like we have to discard another "clock" used by evolutionists to prove an ancient Earth.

We have been taught that oil was produced deep in the Earth about 25,000,000 years or more ago; however, high grade oil has been produced out of cow manure in twenty minutes! And there goes another one!

The Earth can't be five billion years old because the rivers of the world have been constantly carrying sodium to the ocean, and if the world were that old, there would be 100 times more sodium in the ocean than there is!

Studies have been done to chart the volume and rate of sediment accumulation in the Mississippi delta, and it could not be older than 4,000 years![30] The age is found by dividing the weight of sediments deposited yearly into the total weight of the delta. If the Earth were only a few million years old, the Gulf of Mexico would be **full** of sediment!

If helium had been entering the atmosphere at the present rate for five billion years, the atmosphere would contain 200 times as much helium as it now does.

If meteorite dust had been falling upon the Earth at its present rate for over five billion years, there would be a layer of dust around the Earth 182 feet thick! I checked this morning, and it isn't there. This dust has a high concentration of nickel, so it should be found in the crustal rocks; however, it is not found on land or in the oceans. The Earth is young.

The oldest living things on Earth, according to the American Forestry Association, are the Bristlecone pines that grow on the White Mountains of California. They are at least 4,600 years old, no doubt having sprung up from seeds soon after the Flood. Again, an indication of a young Earth. And if the world is ancient, why are there **no trees older than**

## 5,000 years?

There is an average 7 or 8 inches of top soil that sustains all of life on the Earth, while the earth beneath the top soil is as dead as the moon. Scientists tell us that the combination of plants, bacterial decay and erosion will produce six inches of top soil in 5,000 to 20,000 years.[31] If the Earth had been here for 5 billion years, we should have much more top soil than the 7 or 8 inches. It's a young world after all!

The rim of the Niagara Falls is wearing at a known rate each year, and geologists tell us it has taken only 5,000 years to erode from its original precipice.[32]

Newspapers in March of 1980, reported that the sun's diameter appears to have been decreasing by about one tenth percent per century. That means the sun is shrinking about five feet per hour, and that's no problem if you are a creationist. But, you have big trouble if you are an evolutionist!

Dennis Petersen aptly addressed this problem: "If the sun existed only 100,000 years ago it would have been **double its present diameter**. And only twenty million years ago **the surface of the sun would be touching the Earth**."[33] (Emphasis his.) But then we know that didn't happen, don't we? So the Earth and the universe are very young.

We are told that man has been on the Earth one million years, but the population does not reflect that age. There would not be enough room for people to **stand** if the population grew at the same rate as the Jewish people. The Jews started with Jacob about 3,700 years ago, and there are 14,000,000 Jews in the world today. So, assuming 2.4

children per family, and a life-span of 43 years, the world would be **packed** with people if man had been on the Earth a million years.[34]

## ASSUMPTIONS NECESSARY

Make no mistake; evolutionists **must have an ancient** Earth or their house of cards comes tumbling down, so they have brainwashed many people with their "billions of years." They take an assumption and jump to the wrong conclusion. Assumptions **are** necessary in science; however **facts must be added to support or refute an assumption.**

Let me provide an example about assumptions, facts and conclusions: A man leaves home jogging. He goes a little way and turns left; goes a little way and turns left again. Then he goes a little way and turns left again, and arrives home to be met by two masked men. Two questions: Who were the masked men, and why did he leave home? An Innocent One says, "But, there is no way one can know those answers."

However, when you understand that the man was involved in an athletic contest, you revise your preconceptions about home, jogging, masked men, etc. You realize that he was met at home by the umpire and the catcher, and he was jogging because he had hit a home run.[35] Assumptions are necessary, but facts must support them.

Don't swallow the "billions and billions of years." You're being brainwashed with unsupported assumptions. When evolutionists prate on about the ages of the rocks, we creationists will stand firmly on the Rock of Ages.

Fossil trilobites from a stratm in British Columbia.

## CHAPTER TWENTY

# WHAT ABOUT DINOSAURS?

The **fact** of dinosaurs is not in question, the **time** they lived is in question. Dinosaurs were as small as a chicken, and as large as a five story building!

One of the most unusual creatures was the **Pterosaurs**, a flying reptile with huge leathery wings. The smallest fossil of this type was as small as a sparrow and the largest had a wingspan of 51 feet (found in Texas, of course!). Its wings exceeded the 43 feet wingspan of the F-15A jet plane![1]

Sir Richard Owen, British anatomist and paleon-tologist, was the person who gave dinosaurs their name which means, "terrible lizards."

Evolutionists maintain that dinosaurs lived over 65 million years ago, and are a little embarrassed that they seem to have appeared out of nowhere! Dinosaurs flourished for a period then disappeared as quickly as they arrived. Of course, there is no evolution there.

Creationists believe that dinosaurs, in great numbers, lived contemporaneously with men in the same general environment. And according to God's command, they were taken into the ark. (Remember that only two of each basic kind were required, and they could have been young specimens.) The pair of dinosaurs could have been the progenitors of a vast number of dinosaurs following the Flood, or they could have become extinct soon after the Flood. The fossils discovered in recent years could be those creatures that were destroyed in the universal Flood.

Many believe the "dragons" mentioned in the Bible (at least 25 times) were the dinosaurs of paleontology. Scripture indicates the dragons were on the land and in the sea which was true of dinosaurs. Job, who lived soon after the Flood, referred to "behemoth" and "leviathan" in chapters 40 and 41. Behemoth could very well allude to Brontosaurus and leviathan could be a plesiosaur or ichthyosaur.[2]

My Bible notes (not inspired, of course) suggest that behemoth is an elephant; however, Job says that behemoth, moved his tail "like a cedar." Have you ever seen an elephant's tail that could, by any stretch of the imagination (even of an evolutionist's) be compared with a cedar? I think not. Job, shortly after the Flood, I think, referred to dinosaurs.

Leviathan had heavy, close-set scales, had a heart as firm as a stone, and breathed fire! "Come on," says the incredulous one, "you don't believe **that**." Sure I do. I believe that we have fire flies that light up and bombardier beetles that shoot at their enemies from a twin gun in their tails. One thing is sure: leviathan was not a crocodile as some, who want to curry favor with unbelieving scientists, teach. Leviathan, it seems, was a fire-breathing dinosaur that lived in the sea. (Could he be the source of the myths about sea dragons? After all, many myths had a basis in fact.)

Dinosaur fossils have been found on every continent of the Earth, and at times numbering in the thousands. At one time, various dinosaurs roamed the Earth as did the bison in the 1800s in America. Some dinosaurs were meat-eaters while others ate plants.

## DINOSAUR DISAPPEARANCE

Dinosaurs died out, according to evolutionists, about seventy million years **before** man evolved. The quick disappearance of dinosaurs is one of the most troubling questions for the disciples of Darwin. One of the most renown evolutionists, George G. Simpson, admitted this when he stated, "The most puzzling event in the history of life on earth is the change from the Mesozoic, the Age of Reptiles, to the Age of Mammals. It is as if the curtain were rung down suddenly on the stage where all the leading roles were taken by reptiles, especially dinosaurs, in great numbers and bewildering variety, and rose again immediately to reveal the same setting but an entirely new cast, a cast in which the dinosaurs do not appear at all, other reptiles are supernumeraries, and all the leading parts are played by mammals of sorts barely hinted at in the preceding acts."[3]

Simpson is saying that the far ranging dinosaur and other reptiles simply disappeared and their place was taken by mammals. Of course, the problem is much greater than evolutionists usually admit. They must not only find an answer for the Great Dinosaur Disappearance but for other creatures as well. University of California (Berkeley) physicist, Luis Alverez recognized this problem when he said, "The problem is not what killed the dinosaurs but what killed almost all the life at the time."[4]

They don't have the answer, but creationists do. Some evolutionists have speculated (all of them are **expert** speculators) that a disease wiped out all dinosaurs, but that is very unlikely since they ranged the world. It may be possible for a disease to be worldwide, but it is unlikely, especially since there is no evidence for a worldwide plague among dinosaurs.

Some evolutionists have suggested, with a straight face, that dinosaurs may have died of constipation (one of sixty different explanations) because a laxative plant in their diet became extinct. (I kid you not! I believe the evolutionists are mentally constipated.)

In 1946, one paleontologist came up with an original suggestion: There was a global heat wave (of course, no proof) of about two degrees above normal that "baked" the testicles of all male dinosaurs. Of course, that killed any chances for future generations. (Maybe the paleontologist had a baked brain!)

Explanations for the demise of dinosaurs include: that they starved to death; ate poisonous mushrooms; their eggs were eaten by mammals, etc. Scientists would not look nearly so silly if they simply believed what the Bible teaches and what geology suggests--a universal Flood, either immediately or soon thereafter, killed the dinosaurs.

Another suggestion appeared in the January 6, 1982 issue of the **Minneapolis Tribune**: "The extinction of the dinosaurs may have been caused by a giant asteroid that slammed into the earth 65 million years ago. The asteroid collision kicked up a huge cloud of dust containing iridium. The dust obscured the sun for three to six months, destroying the plants on which the dinosaurs fed."[5]

Other scientists have suggested that dinosaurs may have been killed when the temperature dropped quickly;

however, that sets the stage for Noah's Flood which they refuse to consider. There is no way a tropical area will become frigid overnight with only normal (uniformitarian) conditions. Creationists believe that all dinosaurs were destroyed in the Flood. Then there was a worldwide temperature drop, during and following the Flood, that accounts for the demise of any dinosaurs that reproduced after the Flood.

## MEN AND DINOSAURS TOGETHER

Dinosaurs were the most dominant land animals on Earth, and are believed to have lived as early as 225 million years ago then died suddenly, millions of years before man appeared even in ape-like form. However, creationists believe that dinosaurs and men lived contemporaneously. In fact, dinosaur pictographs have been discovered on cave walls in Arizona and Zimbabwe, so where did early men get their knowledge of them if the dinosaurs died out millions of years before men appeared?

Proof of men and dinosaurs living contemporaneously is obviously seen in various places where dinosaur tracks and man tracks have been found together! D. H. Milne and S. Schafersman could see the "handwriting on the wall" in their 1983 article in the **Journal of Geological Education**: "Such an occurrence [dinosaur tracks and man tracks found together] would seriously disrupt conventional interpretations of the biological and geological history and would support the doctrines of creationism and catastrophism."[6]

An example of dinosaurs and mankind's contemporaneous existence is found in Glen Rose, Texas, along the Paluxy River where dinosaur tracks and man tracks have been discovered within inches of each other--uncovered while the press watched the actual discovery! There were weathered dinosaur tracks that led up to an overlay of 12-inch limestone, and the archaeologists believed that the tracks continued under the limestone. While the press looked on, and video cameras whirled, the huge stone was raised revealing that the dinosaur had continued walking along the beach as expected. The tracks continued clearly as if they had been made that morning![7]

No serious scientist questions the authenticity of the dinosaur tracks; however, the man tracks within inches of the dinosaur tracks are questioned by bigoted scientists who must **never** admit that men and dinosaurs lived together. It would ruin their legend.

I participated in a television debate with the head of the Biology Department at a major Midwest university. He had authored two books on evolution and was considered an expert. I pointed out the tracks found at Glen Rose, and challenged his position that dinosaurs and men lived millions of years apart. His explanation for the tracks was that the man tracks had been carved. When he made his "explanation" it was obvious that he was as uncomfortable as a klansman at a NAACP rally. When I pushed him, it was very clear that he did not have current information on the tracks.

The professor must not know that scientists took a human print, sawed it in half, and discovered what one would expect to find if a human had stepped in soft mud: obvious compression underneath the print. Of course, no amount of carving, even by expert carvers, could have produced that result.

The week before the debate, I had returned from the Glen Rose diggings where I was privileged to participate in excavation and managed to uncover a fish fossil. Dr. Carl Baugh, head of the project, has also unearthed trilobites (bottom-dwelling sea creatures) which are supposed to be found at the very bottom of the geologic column! Obviously something was out of place unless the trilobites, the dinosaurs and men all lived at the same time, sharing the same living space. (Gasp!)

Let me try to make it clearer: The archaeologists have unearthed trilobites (the very earliest life forms, according to evolutionists), dinosaur tracks, and man tracks all within inches of each other and all at the same level. Any evolutionist will tell you that the trilobites go at the very bottom of the geologic column, the dinosaurs at the middle, and men at the top, And finding them out of their period is impossible for evolutionists to explain. However, creationists see those facts fitting **perfectly** into the creation model of origins. It is really very simple: The trilobites, dinosaurs and men all lived at the same time in the same general area!

On June 16, 1987, Dr. Baugh also unearthed a fossilized human tooth from a 4 to 8 year-old child within inches of the dinosaur and human tracks! Dr. A. E. Wilder-Smith, holder of three earned Ph.D. degrees flew from his Switzerland home to view the find, and his evolution-shattering response was: "Finding the dinosaur tracks, the trilobites, the man prints, the human tooth all make evolution impossible."[8] And so it does.

But Texas is not the only place where such finds have occurred--much to the consternation of evolutionists. Eighty-six hoof prints of horses were found in rocks in the old Soviet Union that date to the time of the dinosaurs![9] More dinosaur and human prints have also been found together in the former Soviet Union.[10] These have also been found in rock formations in the states of Utah,[11]   Kentucky,[12] and Missouri,[13] and all have been dated by evolutionists at 150-160 million years!

So evolutionists would have us believe that dinosaurs lived more than 70 million years ago, and man first appeared one million years ago; however, the record indicates that both lived at the same time, wiping out evolution somewhat like the dinosaurs were wiped out.

Dinosaur and human prints together cannot be possible if evolution is a fact. But it is **not** a fact. It is a fraud, a farce, a fake and a faith.

# CHAPTER TWENTY-ONE

# WHY ARE EVOLUTIONISTS SO MEAN-SPIRITED?

Evolution, while being conceivable, has not been confirmed and never will be. But most evolutionists are not interested in testing the gradualism of Darwin but rather in protecting their dubious hypothesis (guess). You can imagine the academic chaos if every honest scholar admitted that evolution never took place!

What would it do to every secular college science faculty if creationism were accepted and evolution proved a fraud? They would finally be considered uneducated, unscholarly, and unnecessary. And maybe they would become unemployed! They would fall into the same category as flat-earthers, phrenologists, astrologers and snake handlers! (That is exactly my attitude toward the sooth-sayers of science.)

The museums of the world would have to make some major changes in their displays and propaganda. Their pitch men would have to memorize new facts for their canned speeches.

Public schools would have to retrain their science teachers, and all their science films would have to be destroyed.

Bigoted judges who have ruled in favor of the myth-tellers would feel the sting of public embarrassment and join the ranks of the unemployed. Most likely, they would simply open up shop as attorneys, (and that's what we really need--more attorneys!)

Scientists would no longer seek to confirm the theory of evolution, (now presumed to be a fact **before** beginning an investigation!), and they could get on with the business of true science. Scientists would no longer see only what is "respectable and acceptable," and would look at the evidence, making judgments based on that evidence even if it contradicts popular theory. They would decide that truth is more important than denigrating, denying, and denouncing creationists.

Clergymen and theologians would sheepishly have to apologize to their followers for leading them into the swamps of theistic evolution, day-age theory, gap theory, and other unscriptural nonsense. Such religious leaders might, once again, preach that the Bible, in toto, is reliable as we creationists have been saying for many years.

Publishers of evolutionary tripe would, well, go out of business or start publishing New Age ranting, oriental religious musings, or the sexual fantasies of the latest "pop star."

I can see an assortment of prestigious scientists, clergymen, professors, publishers, film makers, media personalities, and others holding a televised news conference where they apologize profusely to the youth of the world for teaching fraud, falsehood, fakery, and foolishness and calling it "science."

Now, a word of advice: Don't hold your breath for that news conference to take place. It won't, because most evolutionists are vain, venal, and venomous people. They are especially venomous when dealing with or discussing creationists.

Evolutionists, not being very bright and living in a permanent fog, have not understood that creationism has not been disproved, only disbelieved. Evidently they can't make that distinction. Many of them think that making a statement makes it a fact, and by denouncing creationism, they discredit it! Of course, they only discredit themselves.

## PUBLIC DEBATES

Evolutionists have been discrediting themselves for many years, but especially in recent years. And it often takes place in formal public debates with creationists.

The Institute for Creation Research in California has been a pioneer in scientific research, education and apologetics. Dr. Henry Morris and Dr. Duane Gish, with others, have gone into universities and faced evolutionists in public debate--on their turf!

Dr. Morris spoke of those debates: "We are always careful to stick to scientific arguments, especially using the fossil record to show that macro-evolution has not occurred in the past....The evolutionists, however, more often than not, do **not** stick to scientific arguments. They will attack the Bible, show that creationists have religious motivations, argue that one can be religious and still believe in evolution, contend that creationism is not scientific, or attack our personal character or credentials.

"But one thing they will **not** do is give any real scientific evidence for macro-evolution. This is because there isn't any real evidence for macro-evolution! This is why creationists almost always win the debates. We win, not because we are better debaters, but because creation is true, evolution is false, and real science confirms this."[1]

Evolutionists have learned that creationists are not simply Bible thumpers. They expected creationists to quote a few Bible verses then sit down, but evolutionists have discovered that it doesn't work that way. The creationist deals with geology, biology, paleontology, and other branches of science, and they ask questions that evolutionists cannot answer. The creationists also provide a model that proves to be far more acceptable to thinking people than the claptrap of evolution. So, it is getting very difficult to get evolutionists into formal debates.

On November 9, 1987, Dr. Henry Morris was to debate a professor at University of Houston, but a university official cancelled the debate at the last minute because he did not want scientific creationism discussed on his campus. Wait a minute! I thought a university was a place for various ideas to be disseminated, debated and discussed by inquiring minds. Surely it is not a place for propaganda--or is it? Could it be that only what is politically correct is permitted at the University of Houston?

At the National Academy of Sciences all but one person agreed that debating with creationists should be avoided! Well, wonder what they are afraid of? Could it be the truth? If evolution is **really** as scientific as evolutionists claim, they should have nothing to fear. Maybe, just maybe, the evolutionists are not so sure of the correctness of their cause and are embarrassed to face creationists in formal debate.

One of the best known evolutionists, Niles Eldredge, refers to these debates in his book, **Monkey Business**: "Creationists travel all over the United States, visiting college campuses and staging 'debates' with biologists, geologists, and anthropologists. The creationists nearly always win. The audience is frequently loaded with the already converted and the faithful. And scientists, until recently, have been showing up at the debates ill-prepared for what awaits them. Thinking the creationists are uneducated, Bible-thumping clods, they are soon routed by a steady onslaught of direct attacks on a wide variety of scientific topics. No scientist has an expert's grasp of all the relevant points of astronomy, physics, chemistry, biology, geology, and anthropology. Creationists today--at least the majority of their spokesmen--are highly educated, intelligent people. Skilled debaters, they have always done their homework. And they nearly always seem better informed than their opponents, who are **reduced too often to a bewildered state of incoherence.**"[2] (Emphasis added.)

Well, that **is** some admission from a leading evolutionist! But note that he tried to remove some of the sting by charging creationists with loading the audience with our own people. That is simply not true. In fact, the opposite is true. **Every** university debate I have been involved with has proved Eldredge wrong. In fact, I have never known of more than two or three people who agreed with me out of many hundreds who attended the debates.

The same is true of talkshow audiences. Whether it was the Morton Downey, Jr. show where I debated a porn king, the Jerry Springer Show where I debated four homosexuals or other shows, the opposing crowd is always there in force. During a televised debate in Tampa with the area's leading homosexual, the first row was loaded with 19 homosexuals who jeered when I quoted the experts on homosexuality or Bible principles. As can be expected, they cheered when my opponent praised his perverted life style. (I called it their "death style.") No, Eldredge was wrong. We creationists do not "load" the audience (not that I'm above such a thing). It's only that it can't be done in a secular university.

The unfair, unreasonable, unacceptable, unblushing, unbecoming, uncivil, unconscionable and ungentlemanly tactics of evolutionists are obvious to all who have seen them in debates or read their books and articles. (Leading evolutionist Stephen J. Gould calls creationists, "kooky," "yahoos" and "latter-day antediluvians" and he characterizes scientific creationism as "the nonsense term of the century.") However, Steve refuses to make those charges in public debate where informed creationists can make him look like a "yahoo." He recently declared, "Creationist-bashing is a noble and necessary pursuit these days."[3] Steve hates creationists because he obviously hates our God.

Like Gould, H. L. Mencken was a writer with bile problems and a hatred toward creationists. He used the Scopes Trial as a showcase for his vindictive, vicious and vitriolic journalism. He gave "thanksgiving" that creationist William Jennings Bryan had not been elected President of the United States, although he was not thrilled with Coolidge. After ridiculing Coolidge, he referred to the Scopes Trial and ridiculed Bryan and Christians everywhere. He suggested that we believe the Earth is square (we don't, but some of us believe his head was), witches should be put to death and

Jonah swallowed the whale![4] Mencken was a hater of God and of creationists.

Evolutionists hate creationists and anyone who believes that the Bible is the authoritative Word of God. They are as uncomfortable as a dog in hot ashes when we talk of personal accountability to a sovereign God. They are bigots and haters with a few exceptions.

## UNFAIR TREATMENT

Dr. Alfred Rehwinkel summed it up very well: "Meanwhile, their [evolutionists] unproven theories will continue to be accepted by the learned and the illiterate alike as absolute truth, and will be defended with a fanatic intolerance that has a parallel only in the bigotry of the darkest Middle Ages. If one does not accept evolution as an infallible dogma, implicitly and without question, one is regarded as an unenlightened ignoramus or is merely ignored as an obscurantist or a naive, uncritical fundamentalist. Such a man will have difficulties finding a publisher for a manuscript presenting a view contrary to this theory, and if published his book will not be recommended for a textbook in schools or colleges. It will not be easy for him to secure a position in the science department of one of our leading universities. Anyone who does not join in the modern 'Baal-worship' is a heretic and is treated accordingly."[5]

Rehwinkel is correct. Talk to professors in various universities who have "come out of the closet" and have been scorned, ridiculed, and even fired for exercising their academic freedom. Talk to those who have been refused a Ph.D. because they rejected evolution. Those same university officials speak of pluralism, fairness, and diversity, and it's a wonder they don't gag on their words.

Dr. John Patterson of Iowa State University delivered a speech before the Iowa Academy of Sciences in which he suggested that creationists should lose their faculty contracts and even their degrees![6] What a bigot, but then bigots are as easy to find in a university as a bowling ball in a bathtub.

Is it any wonder that some scientists are fearful of questioning Darwin when bigots such as Patterson are in positions of power?

Another example is found in **Nature**, written by A. T. J. Hayward: "There are more anti-Darwinists in British universities than you seem to realize. Among them is a friend of mine who holds a chair in a department of pure science....If his friends ask why he keeps quiet about his unorthodox views, he replies in words like those used recently in another connection by Professor Ian Roxburgh: '...There is a powerful establishment and a belief system. There are power seekers and career men, and if someone challenges the establishment he should not expect a sympathetic hearing.'...The majority of biologists accept the prevailing view uncritically--just as a great many competent Russian biologists were once brainwashed into accepting Lysenko's quackery. Others have thought for themselves and came to realize the flaws in contemporary Darwinism. But for them to speak out would be to invite ridicule and would probably ruin their careers."[7] There is an example of prejudice so strong that truth and scholarship are the victims, but **none dare call it censorship.**

Dr. Henry Morris knew of the mistreatment of creationists at Virginia Tech: "I have known students who have failed courses, and some who have been denied admission to graduate school or have been hindered from obtaining their degrees, largely for this very reason. When I was on the faculty of Virginia Tech, a professor who was on the graduate faculty there in the Biology Department told me that he would never approve a Ph.D. degree for any student known to be a creationist in his department, even if that student made straight A's in all his courses, turned in an outstanding research dissertation for his Ph.D., and was thoroughly familiar with all the evidence for evolution."[8]

A good example of the above took place in 1964 at the University of Arizona where Clifford Burdick, a creationist, was doing doctoral research, and some of his discoveries conflicted with the religion of evolution. When his work came to the attention of the Powers That Be at the university, he was refused his doctor's degree.[9] Wonder what happened to academic freedom? Forget it if you are a creationist. The unwritten university law is, you can research **how** evolution happened but not **whether** it happened!

Do you wonder why evolutionists are so bitter, mean and cruel toward creationists? They hate our God. They know we are a challenge to their fairy tale (and their jobs), and they know creationism has better answers to the questions of where life came from and how it got here than does evolution. They prefer to ignore us rather than to look at the evidence because they are afraid of what they might see. Their minds are as closed as a miser's wallet.

One honest evolutionist, Dr. W. Scot Morrow, alluded to this attitude when he wrote the foreword for **Creation's Tiny Mystery** by Robert Gentry: "If I were to follow the unwritten, but commonly understood guidelines laid down by my fellow evolutionists, many of whom are agnostics like myself, when presented with a book written by a fundamentalist Christian on the topic of 'creation,' I would ignore the work. Of course, I might kick over the traces a bit, skim through the thing quickly--one MUST be fair, you know--and then give the document a decent quiet burial in the nearest wastebasket. After all, those among us who have brains in our head instead of rocks...know that (1) science and religion are immiscible, (2) true scientists cannot be creationists, (3) creationists cannot be scientific, let alone scientists, (4) the last factor is doubled and redoubled--in spades--for fundamentalists."[10]

Morrow takes a deft swat at his bigoted colleagues with a not-too-subtle use of sarcasm. He is telling us what most evolutionists would do with a book written by a creationist. They would not treat it seriously nor fairly nor honestly. It sounds as if he is saying that they are biased, bogus and bigoted. But **surely** not.

Gentry built a case for a younger Earth (than demanded by evolutionists) in **Creation's Tiny Mystery** which resulted in the loss of his research grants and his job-- thanks to the bigots and champions of academic freedom!

Frank Manheim was a Harvard undergraduate who proved the hypocrisy of a small-minded college professor. He was doing B+ work and it came time for his term paper in a biology course. His professor had hammered on two themes: evolution and challenging authority, so Frank took his prof at his word. He wrote a paper on evolution, challenging most scientific authorities!

When Frank received his paper, he was astonished to see a D- grade. He approached the professor for an explanation for the bad grade. He explained that he had only followed the professor's guidance in challenging authority at the highest level. And, in passing, Frank mentioned that he didn't personally doubt evolution!

The professor answered, "Oh, you don't?" Very pleased that his student had not apostatized to creationism, he changed his grade on the spot to a B. (Such a professor should be suspect even when he quotes the multiplication tables.)

Frank wrote, "...remembering the hypocrisy of my early professor and the fallibility of many noted scientists, I have been chagrined by the arrogance and vituperation with which the creationists have been treated. In fact, I would argue that intolerance and absolutism--no matter how sure we are of our case--are not really consistent with the scientific method and give science a bad image."[11] Right on, Frank. Another honest evolutionist.

Evolutionists will use any opportunity to ridicule creationists, but they will not deal with them. Most of us have been ridiculed by professors at secular universities because of our faith in the Word of God.

J. W. Lane has observed the unfairnes of professors and reports, "I have myself sat in class after class in the sciences and humanities in which any idea remotely religious was belittled, attacked, and shouted down in the most unscientific and emotionally cruel way. I have seen young students raised according to fundamentalist doctrine treated like loathsome alley cats, emotionally torn apart, and I never thought that this sort of treatment was any better than the treatment that religious prelates, who held authority, gave Galileo. Why scream about the inhumanity of nuclear war if you are also willing to force people of fundamentalist faiths to attend public schools in which their most cherished beliefs will be systematically held up to ridicule and the young children with it....Do you call it fair?"[12]

Of course, it isn't fair, but then most evolutionists are dishonest people since they refuse to look at both sides of the issue. Even Darwin was desirous of scientists looking at both sides to come to a conclusion. He wrote: "I look with confidence to the future--to young and rising naturalists, who will be able to view both sides of the question with impartiality...."[13]

But most evolutionists are unfair, unkind, unscholarly and unreasonable, and are not willing to "view both sides of the question with impartiality." After a life-changing experience with Jesus Christ, that will change. Of course, an evolutionist can be fair, kind, scholarly, reasonable and **wrong** without knowing Christ as Savior, but when a person is born from above, he will be honest with himself and with others, and his attitude, aspirations, and his assumptions will reflect a biblical life view.

Most evolutionists are unbelievers, although a few talk vaguely about an ambiguous God somewhere. Many are atheists who cling to the religion of evolution to support their unbelief and their free-wheeling life style. Remember this principle: If evolution is a reality, then death was the norm for millions of years long before man "evolved." That would mean that death was not the result of Adam's sin, and sin is a fiction. And, we don't need a Savior if we aren't sinners.

Evolutionists and atheists are fully aware of that fact. One wrote, "Christianity has fought, still fights, and will fight science to the desperate end over evolution, because evolution destroys utterly and finally the very reason Jesus' earthly life was supposedly made necessary. Destroy Adam and Eve and the original sin, and in the rubble you will find the sorry remains of the son of God. Take away the meaning of his death. If Jesus was not the redeemer who died for our sins, and **this is what evolution means**, then Christianity is nothing!"[14]  (Emphasis added.) Well, he said it well, didn't he? And it reveals why they are so adamantly against us and what we believe.

**However, when an evolutionist meets Christ, he will lose his mean streak because he will become a new person in Him.**

Iron Hammer Found Incased in Stone

# EPILOGUE

Well, that's it. You have read the "other" side of the evolution/creation debate. You will admit, at least, that you never heard that position in public schools. Maybe you can see that we Christians have an impressive amount of science to support our position. It is not, as evolutionists say, simply a matter of quoting a few Bible verses.

It is very interesting that the general public supports teaching creation in public schools rather than evolution alone. An AP-NBC News poll reported 86% to 8% favored such teaching,[1] but don't hold your breath waiting for it to happen unless you look good in blue!

British paleontologist Dr. Colin Patterson suggests that evolution should **not be taught** in public high schools, (and no honest, informed, fair-minded person would teach it **as a fact**). Patterson began asking scientists to tell him "one thing" they **knew** about evolution. While lecturing to biologists at the American Museum of Natural History in New York City, he said, "I tried that question on the geology staff at the Field Museum of Natural History and the only answer I got was silence. I tried it on the members of the Evolutionary Morphology Seminar in the University of Chicago, a very prestigious body of evolutionists, and all I got there was silence for a long time and eventually one person said, 'I do know one thing--it ought not to be taught in high school.' "[2]

Well, even a Harvard professor should be able to understand that simple principle but don't bet the farm on that. Liberals talk about fairness, understanding, equal rights, diversity, but they don't practice what they teach. They are the most inconsistent people in America and Canada. Note the following examples from three evolutionists.

Clarence Darrow, lawyer for John Scopes (paid by the ACLU), whined that it was unfair and un-American that the school children of Tennessee only hear one theory of origins (creation). They should hear about evolution, and let them make up their own minds. After all, ideas can't hurt you. That was his general pitch, but then they were on the outside looking in. Now it is different. The ACLU and other radicals have taken over the public school system, but of course, it is still funded by us taxpayers (who have little voice in our school system).

D. F. Malone, another attorney for John Scopes at the Dayton, Tennessee trial argued, "For God's sake, let the children have their minds kept open--close no doors to their knowledge; shut no door from them...."[3]

Finally, John Scopes himself said that "[E]ducation, you know, means broadening, advancing, and if you limit a teacher to only one side of anything the whole country will eventually have only one thought....I believe in teaching every aspect of every problem or theory."[4]

Well, things do change, don't they? It is now illegal to teach the truth of man's origins to children, although 86% of Americans want it done. Maybe Christian teachers (and fair-minded non-Christian teachers) in the public schools should teach creationism illegally! And accept the consequences that follow!

Maybe we need to be more militant and take our "religion" outside the church walls into the public sector where we can have more impact. In fact, how can a principled teacher do otherwise? Surely they can't justify teaching error. After all, creation is the truth and is supported by scientific fact.

You will be held accountable for what you have read on these pages! Maybe you hadn't planned on being more responsible simply by reading this book, but you are.

You will now ask evolutionists to explain as to when matter, space and time first originated--and demand an answer.

You will expect an explanation as to when energy originated. Since the First Law of Thermodynamics requires that energy is not now being formed or destroyed--it has always remained static--then it is a corollary that no natural process can create energy, it must have been originally created by a supernatural Force.

You will demand an answer as to why the moon is not covered in dust a mile deep if it is almost five billion years old.

You will want to know why we have seen hundreds of stars "die" but not one has been "born." You will demand an answer.

You will demand a reasonable answer as to how life first started, and you won't accept the cockamamie idea of spontaneous generation.

You will expect evolutionists to explain why not one fossil has been found (out of billions) that is transitional from one species to another.

You will ask why coal is not being formed anywhere

in the world as it has in the past. What has happened to change its formation except the absence of a second universal Flood?

You will expect them to explain why no fossils are being formed today as in the past. After all, wouldn't uniformitarianism demand that?

You will require an answer as to why every major mountain range on Earth is rich in fossilized sea life. I've heard of flying fish but that's ridiculous.

You will also want to know why evolution has suddenly stopped after millions of years.

You will also ask where and how man acquired morality, his awareness of beauty, conscience, the ability to blush, weep and laugh, his feelings of guilt, ability to repent, etc.

You will ask the evolutionists how, when, and where sex originated and how a male and female evolved over millions of years yet arrived at sexual maturity at the same time in the same location! This would be true with **all** creatures! And please don't sneer at them when they answer. Maybe a giggle or two would be appropriate.

You will ask those who admit that the gradualism of Darwin did not happen and who have accepted the reincarnation of the hopeful monster theory (punctuated equilibrium) if they really believe such nonsense. You will, of course, try not to fall to the ground, holding your sides in laughter when they inform you they are serious. A simple smirk will do.

You now realize that scientists have been wrong many times, and evolutionists have made a career of it! You realize that the evolutionist is sure, smooth and superior--and wrong while you are sure, smooth and superior--and right! But don't gloat, after all, we are all ignorant in some areas.

You and I dislike the evolutionist's corrupting of young people, but you must realize that his eyes are blind and his mind is darkened. He has been taken captive according to the principles of the world rather than according to Christ. So you will pity him and pray for him and per chance he will not only accept creationism but accept Christ as personal Savior as well.

One thing is sure: The evolutionist must deal with God, either here or hereafter. What a tragedy to live by his faith in reason (in the religion of evolution), so sure he is right, grasping his Doctorate of Pomposity, only to realize his life is a nightmare. He has pulled himself up the mountain of learning, but still in ignorance  he approaches the highest peak; and as he drags himself over the final rock, he is greeted by Jehovah God, sitting on a Throne over which is written, "In the beginning, God created the heavens and the earth."

# REFERENCES

## FOREWORD

1. Niles Eldredge, *The Monkey Business: A Scientist Looks at Creationism*, p. 24, 1982.
2. S. Lovtrup, *Darwinism: The Refutation of a Myth*, p. 422, 1987.

## INTRODUCTION

1. Bernard Stonehouse, Introduction to M. Pitman, *Adam and Evolution*, p. 12, 1984.
2. T. H. Huxley, *Life and Letters of Thomas Henry Huxley*, v. I (Ed. L. Huxley), p. 254, 1903.
3. Quoted in M. Bowden, *The Rise of the Evolution Fraud*, p. 214, 1982.
4. G. de Beer, *A Handbook of Evolution*, 1958, quoted in *The Rise of the Evolution Fraud*, p. 214, 1982.
5. N. Macbeth, *Darwin Retried*, p. 134, 1971.
6. A. Keith, quoted in *Science and Scripture*, p. 17, 1971.
7. M. Denton, *Evolution: A Theory in Crises,* p. 327, 1986.

## CHAPTER ONE

1. M. Denton, *Evolution: A Theory in Crisis*, p. 75, 1986.

2.   *The World Book Dictionary,* Doubleday and Co., p. 1846, 1974.
3.   Duane Gish, *Evolution: The Challenge of the Fossil Record,* p. 12, 1985.
4.   G. G. Simpson, *Science*, v. 143, p. 769,  1964.
5.   N. Macbeth, *American Biology Teacher*, p. 496,  1976.
6.   L. C. Birch and P. R. Ehrlich, *Nature,*  v. 214, p.349, 1967.
7.   Duane Gish, *Evolution: The Challenge of the Fossil Record,* p. 252, 1985.
8.   Ibid.
9.   C. Singer, *A History of Biology*, p. 487, 1931.
10.   Walter J. Bock,  "Evolution by Orderly Law," *Science*, v. 164, pp. 684-685, May 9, 1969.
11.   G. de Beer,  *An Atlas of Evolution*, p. 159, 1964.
12.   Robinson, Haeckel, Ernst, 5 *Encyclopaedia Britannica: Miropaedia* p. 610, 15th ed. 1976.
13.   R. Danson, "Evolution," *New Scientist*, v. 29, p. 35, 1971.
14.   W. R. Thompson, Introduction, to *The Origin* by C. Darwin,  p. 12, 1956.
15.   K. Thomson, "Ontogenty and Phylogeny Recapitulated," *American Scientist*, v. 76, p. 273,  May-June 1988.
16.   R. Carrington,  *The Story of our Earth*, p. 64, 1956.
17.   Carl Sagan, quoted by Ken Ham in ICR, no. 48, Dec. 1992 from *Parade Magazine*, April 22, 1990.
18.   Quoted in *Evolution?,* J. W. G. Johnson, p. 20, 1986.
19.   S. R. Scadding, "Do 'Vestigical Organs' Provide Evidence for Evolution?" *Evolutionary Theory,* v. 5, p. 173, May, 1981.
20.   Phillip Johnson, *Darwin on Trial,* p. 73, 1991.

21. T. H. Huxley, *Life and Letters of Thomas Huxley*, v. 1, (L. Huxley, ed.), p. 241, 1903.
22. R. D. Alexander, in *Evolution versus Creationism: The Public Education Controversy*, p. 91, 1983.
23. C. Darwin, *The Origin of Species*, p. 482, repr. 1964.
24. D. M. Watson, *Nature*, v. 124, p. 233, 1929.
25. S. Lovetrup, *Darwinism: The Refutation of a Myth*, p. 422, 1987.
26. Dobzhansky, Book Review of *The Evolution of Living Organisms*, p. 376, 1975.
27. B. Leith, *The Descent of Darwin: A Handbook of Doubts about Darwinism*, p. 11, 1982.
28. Stephen Stanley, *Macroevolution: Pattern and Process*, p. 39, 1979.
29. P. Lemoine, Introduction: De L'Evolution? in 5 *Encyclopedie Francaise*. p. 6, P. Lemoine ed., 1937.
30. Hsu, "Darwin's Three Mistakes," *Geology*, v.14, p. 534, J 1986.
31. N. Macbeth, *Darwin Retried*, pp. 77-78, 1971.
32. A. Koestler, *Janus: A Summing Up*, p. 192, 1978.

## CHAPTER TWO

1. Quoted in *Evolution?*, J. W. G. Johnson, p. 2, 1986.
2. Ibid., p. 3.
3. A. Thompson, *Biology, Zoology, and Genetics: Evolution Model vs. Creation Model*, p. 76, 1983.
4. A. Clark, "Animal Evolution," *Quarterly Rev. Biology*, v. 3, p. 539, 1928.
5. H. S. Lipson, "A Physicist Looks at Evolution," *Physics Bulletin*, p. 138, May, 1980.
6. D. M. S. Watson, *Nature*, v. 124, p. 233, 1929.

7. Edward Blick, *Correlation of the Bible and Science*, p.2, 1976.
8. Albert Sippert, *From Eternity to Eternity*, p. 55, 1989.
9. C. S. Smith, "Materials and the Development of Civilization and Science," *Science,* v. 148, p.908, May 14, 1965.
10. Ibid., p. 674.
11. H. Morris, *The Biblical Basis for Modern Science*, p. 446, 1984.
12. Ibid.
13. Quoted in *Origin Science*, N. Geisler and J. K. Anderson, p. 122, 1987.
14. L. John, *Cosmology Now*, p. 92, 1976.
15. Joseph Silk, *The Big Bang*, quoted in *Eternity to Eternity* by Sippert, p. 154, 1989.
16. M. Bishop, B. Sutheland, and P. Lewis, *Focus on Earth Science*, p. 470, Teacher's Ed., 1981.
17. J. Trefi, *The Moment of Creation: Big Bang Physics* p. 48, 1983.
18. Leslie, "Cosmology, Probability, and the Need to Explain Life," in *Scientific Explanation and Understanding,* p. 54, N. Rescher, ed., 1983.
19. F. Hoyle, *The Intelligent Universe: A New View of Creation and Evolution,* p. 185, 1983.
20. P.Davies, "Universe in Reverse--Can Time Run Backwards?" *Second Look,* p. 27, Sept. 1979.
21. W. T. Brown, Jr., *In the Beginning*, p. 11, 1989.
22. Davies, "Chance or Choice: Is the Universe an Accident?," *New Scientist,* v. 80, p. 506, 1978.
23. An address by M. Planck (May 1937), repr. in A. Barth, *The Creation,* p. 144, 1968.

24. A. Einstein, *Letters A' Maurice Solovine,* p. 115, 1956.
25. F. Hoyle, *The Intelligent Universe: A New View of Creation and Evolution* p. 186, 1983.
26. W. Bird, *The Origin of Species Revisited,* p. 437, 1989.
27. A. Krauskopt and A. Beiser, *The Physical Universe,* p. 645, 3rd ed., 1973.
28. H. Shapley, *On the Evolution of Atoms, Stars and Galaxies* in Adventures in Earth History, p. 79, P. Cloud ed., 1970.

## CHAPTER THREE

1. William Provine, "Evolution and the Foundation of Ethics," *MBL Science,* v. 3, no. 1, p. 29, quoted in *Darwin on Trial,* p. 124, 1991.
2. John R. Rice, *Genesis,* p 40, 1975.
3. A. Huxley, *Ends and Means,* pp.312-315, 316, 1937.
4. Phillip Johnson, *Darwin on Trial,* p. 114, 1991.
5. George Wald, "The Origin of Life," in *The Physics and Chemistry of Life,* by the editors of *Scientific American,* p. 12, 1955.
6. Personal letter, James Barr, April 23, 1984.
7. Harry Rimmer, *Modern Science and the Genesis Record,* pp. 21-22, 1937.
8. Donald Chittick, *The Controversy,* pp. 107-108, 1984.

## CHAPTER FOUR

1. Quoted in C. Wilson, *Creation or Evolution,* p. 90, 1985.
2. Ibid.
3. William Albright, *Archaeology and the Religions of*

*Israel*, p. 176, 1956.
4.  William Albright, *The Archaeology of Palestine*, Rev.
      ed. Harmondsworth, p. 127-128, 1960.
5.  William Albright, *The Biblical Period from Abraham to
      Ezra*, 1960.
6.  M. Burrows, *What Mean These Stones?*, p. 291, 1956.
7.  Merrill Unger, *Archaeology and the New Testament*,
      pp. 1-2, 1962.
8.  Address by R. G. Lee, in *Maranatha*, p. 6, May 1989.
9.  Edward Blick, *Correlation of the Bible and Science*, p.
      3, 1976.
10. Ibid.
11. M. M. Waldrop, "Delving into the Hole in Space,"
      *Science* v. 214, p. 1016, Nov. 21, 1981.
12. E. Blick, *Correlation of the Bible and Science*, p. 2,
      1976.
13. Robert Jastrow, *Christianity Today*, p. 15, Aug. 6,
      1982.

## CHAPTER FIVE

1.  Robert Jastrow, "God's Creation," *Science Digest*,
      Special Spring Issue, p. 68, 1980.
2.  Donald Chittick, *The Controversy*, p. 55, 1984.
3.  M. Grene, *Encounter*, p. 49, Nov. 1959.
4.  C. Patterson, *The Listener*, Oct. 8, 1981.
5.  H. Nilsson, *Synthetische Artbildung*, pp. 1185,
      1212, 1953.
6.  B. Leith, *The Descent of Darwin: A Handbook of
      Doubts about Darwinism*, p. 11, 1982.
7.  J. Huxley, *Religion Without Revelation*, 1957, p. 181,

8.  *Christian Science Monitor*, Jan. 4, 1962.
9.  G. A. Kerkut, *Implications of Evolution*, p. 150, 1960.
10. T. Chardin, as cited in Francisco Ayola, "Nothing in Biology Makes Sense Except in the Light of Evolution: Theodosius Dobzhansky, 1900-75," *Journal of Heredity*, v. 68, no. 3, p. 3, 1977.
11. Julian Huxley in *Issues in Evolution,* v. 3 of *Evolution After Darwin,* S. Tax ed., p. 260, 1960.
12. J. Huxley, and J. Bronowski, *Growth of Ideas*, p. 99, 1986.
13. Cited in *Moody Monthly*, T. Woodward, p. 20, Sept. 1988.
14. Scott  Huse, *The Collapse of Evolution,* p.3, 1988.
15. G. Simpson, *Science*, v. 143, p. 769, 1968.
16. L.  Matthews, Introduction to *The Origin of Species*, rep. by J. M. Dent and Sons, London, p. X, 1971.
17. Ibid. XI.
18. W. Smith, *Teilhardism and the New Religion*, p. 242, 1988.
19. Quoted in *Creation Scientists Answer Their Critics* by D. Gish, p. 64, 1993.
20. Quoted in *The Neck of the Giraffe,* F. Hitching, p. 89, 1983.
21. Quoted in *Ammunition,* by N. Sharbaugh, p. 25, 1991.
22. D. H. Kenyon, "The Creationist View of Biological Origins," *NEXA Journal*, p. 33, Spring 1984.
23. H. Lipson, "A Physicist Looks at Evolution," *Physics Bulletin,* p. 138, May 1980.
24. J. Mill, *Three Essays on Religion*, p. 174, 2nd ed. 1814, rep. 1969..
25. F. Hoyle, "The Universe: Past and Present Reflections,"

*Engineering and Science*,  v. 12, Nov. 1981.
26.  W. Bird, *The Origin of Species Revisited*,  v. 1, p. 302, 1989.
27.  S. Huse, *The Collapse of Evolution,*  p. 6, 1988.

### CHAPTER SIX

1.  Quoted in *The Rise of the Evolution Fraud*, M. Bowden, p. 39, 1982..
2.  F. Darwin,  ed., *The Life and Letters of Charles Darwin*, 3 vols., John Murray, London, v. 1,  p. 45, 1988.
3.  Ibid., v. 1,  p. 307.
4.  M. Denton, *Evolution: A Theory In Crisis*,  p. 69, 1986.
5.  Ibid., p. 104.
6.  S. J. Gould, "Eternal Metaphors of Palaeontology" in *Patterns of Evolution*, ed. A. Hallam,  p. 7, 1977.
7.  L. Eisley,  *Darwin's Century*,  p. 242, 1959.
8.  Paul Kildare, "Monkey Business," *Christian Order,* p. 591, Dec. 1982.
9.  Ibid., p. 592.
10. Daniel Gasman, *The Scientific Origins of National Socialism: Social Darwinism in Ernst Haeckel and the German Monist League*, p.168, 1971.
11.  Edward Simon, "Another Side to the Evolution Problem," *Jewish Press*, p. 24-B, Jan. 7, 1983.
12.  Arthur Keith, *Evolution and Ethics*,  p. 28, 1947.
13.  Charles Darwin, *The Descent of Man*, pp. 241-242, 1901.
14.  *Huxley's Lectures and Lay Sermons*,  p. 115, 1865.
15.  Henry F. Osburn, "The Evolution of Human Races,"

*Natural History*, p. 129, April 1980,--reprinted Jan/Feb. 1926 issue.

16. Edwin Conklin, *The Direction of Human Evolution*, p. 34, 1921.
17. Ibid. p. 53.
18. M. Bowden, *The Rise of the Evolution Fraud*, pp. 188-189, 1982.

**CHAPTER SEVEN**

1. G. G. Simpson, *Science*, v. 143, p. 769, 1964.
2. G. A. Kerkut, *Implications of Evolution*, p. 157, 1960.
3. "Review of Kerkut's Book," *American Scientist*, v. 49, p. 240, June, 1961.
4. T. Dobzhansky, *Evolution*, v. 29 , p. 376, 1975.
5. Ibid.
6. P. Grasse, *The Evolution of Living Organisms*, p. 6, trans. 1977.
7. Paul Lemoine, Introduction: De L'Evolution?, in 5 *Encyclopedie Francaise*, p, 6, P. Lemoine ed., 1937.
8. S. Lovtrup, *Darwinism: The Refutation of a Myth*, p. 352, 1987.
9. Stephen Stanley, *Macroevolution: Pattern and Process*, p. 39, 1979.
10. H. Lipson, "A Physicist Looks at Evolution," v. 31 *Physics Bulletin,* p. 138, 1980.
11. Michael Denton, *Evolution: A Theory in Crisis*, p. 75, 1986.
12. Edwin Conklin, *Readers Digest*, Jan. 1963.
13. Cited in *Correlation of the Bible and Science*, Edward

Blick, p. 10, 1976.

14. Ibid.
15. Ibid.
16. Heribert Nilsson, *Synthetische Artbildung,* pp. 1185, 1212. 1953.
17. Quoted in *Evolution?* by J. Johnson, p. 3. 1986.
18. Quoted in *Evolution: The Incredible Hoax* by G. Lindsay, p. 16, 1977.
19. Zygmunt Litynski, "Should We Burn Darwin?" *Science Digest*, v. 51, p. 61, Jan. 1961.
20. W. R. Bird, *The Origin of Species Revisited*, v. 1, p. 146, 1989.
21. B. B. Vance, and D. F. Miller, *Biology For You*, p. 580, 1950.
22. J. M. Savage, *Evolution*, 1965.
23. R. B. Goldschmidt, *American Science*, v. 45, p. 388, 1957.
24. J. Huxley, *Issues in Evolution,*, v. 3 of *Evolution After Darwin,* S. Tax, ed., p. 41, 1960.
25. M. J. Kenny, *Teach Yourself Evolution*, p. 159, 1966.

## CHAPTER EIGHT

1. Remarks taken from *Applied Christianity*, p. 8, May 1974.
2. "At Random: A Television Preview," in *Issues in Evolution,* v. 3 of *Evolution after Darwin*, p. 41, 1960.
3. David Hume, *Letters of David Hume*, 2 vols., p. 187, 1932.
4. Robert Jastrow, *God and the Astronomers*, p. 14, 1978.

5. Ibid., p. 15.
6. Ibid., p. 18.
7. Cressy Morrison, *Readers Digest*, Dec. 1946.
8. Cited in *From Eternity to Eternity*, Albert Sippert, p. 167, 1989.
9. Albert Sippert, *From Eternity to Eternity*, p. 175, 1989.
10. Scott Huse, *The Collapse of Evolution*, p. 70, 1988.
11. Carl Sagan, *Cosmos*, p. 278, 1980.
12. Michael Denton, *Evolution: Theory in Crisis*, p. 330, 1986.
13. *Science News Letter*, p. 148, Sept. 5, 1955.
14. Carl Sagan, "Life," *Encyclopaedia Britannica: Macropaedia* p. 894, 15th ed. 1974.
15. G. G. Simpson, *The Meaning of Evolution*, p. 16, 1949.
16. Donald Chittick, *The Controversy*, p. 67 1984.
17. Walter Brown, Jr., *In the Beginning*, p. iii., 1989.
18. 2 *The Life and Letters of Charles Darwin*, p. 67, F. Darwin ed., 1887.
19. Charles Darwin, *The Origin of Species*, pp. 186-187 (First ed. 1889, repr. 1964).
20. Scott Huse, *The Collapse of Evolution*, p. 71, 1988.
21. G. Hardin, *Nature and Man's Fate*, p. 72, 1961.
22. David Menton, *Wall Street Journal*, July 15, 1979.
23. Frank Salisbury, "Doubts About the Modern Synthetic Theory of Evolution," *American Biology Teacher*, pp. 336-338, Sept. 1971.
24. Allan Sandage, "A Scientist reflects on religious belief," *Truth*, v. 1, p. 54, 1985.
25. *God In Creation*, Moody Press, Chicago, p. 56, 1982.
26. R. E. Kafahl, *Handy Dandy Evolution Refuter*, p. 27,

1980.

27. M. Pitman, *Adam and Evolution*, p. 194. 1987.

28. R. E. Kafahl, *Handy Dandy Evolution Refuter*, p. 159, 1980.

## CHAPTER NINE

1. *Encyclopaedia Britannica*, v. 4 Micropedia, p. 623.

2. D. Gish, *Evolution: The Fossils Say No!*, p. 122, 1976.

3. C. Darwin, (1872) *The Origin of Species*, 6th ed., 307, 1962.

4. D. B. Kitts, "Paleontology and Evolutionary Theory," *Evolution*, v. 28, p. 467, Sept. 1974.

5. D. M. Raup, *Field Museum of Natural History Bull.*, v. 50, p. 22, 1979.

6. T. N. George, *Science Progress*, v. 48, p. 1, 1960.

7. A. Zihlman and H. Lowenstein, "False Start of the Human Parade," *Natural History,* v. 88, p. 86, Aug.-Sept. 1979.

8. David Pilbeam, "Rearranging Our Family Tree," *Human Nature*, p. 45, June 1978.

9. P. E. Johnson, *Darwin on Trial*, p. 59, 1991.

10. Solly Zuckerman, *Beyond the Ivory Tower*, p. 64, 1970.

11. Ibid., p. 19.

12. John Pfeiffer, *The Emergence of Man*, p. 10, 1969.

13. Solly Zuckerman, *Beyond the Ivory Tower*, p. 64, 1970.

14. Charles Darwin, *My Life and Letters*, v. 1, p. 210, cited in *The Rise of the Evolution Fraud*, p. 214.

15. Cited by W. Bird in *The Origin of Species Revisted*, v. 1, p. 44, 1989.

16. S. J. Gould, *The Panda's Thumb*, p. 188, 1982.

17. *The Guardian Weekly*, v. 119, no. 22, Nov. 26, 1978.
18. *The Guardian*, Nov. 21, 1978 quoted by Bowden in *The Rise of the Evolution Fraud*, p. 217, 1982.
19. W. E. Swinton, in *Biology and Comparative Physiology of Birds*, v. 1, A. J. Marshall, ed., p. 1, 1960.
20. Henry Morris, *The Biblical Basis for Modern Science*, p. 341, 1984.
21. J. L. Marx, *Science*, p. 284, 1978.
22. Quoted in *Evolution?* by J. Johnson, p. 33, 1986.
23. Duane Gish, *Evolution: The Challenge of the Fossil Record*, p. 114, 1985.
24. Michael Denton, *Evolution: A Theory in Crisis*, p. 176, 1986.
25. S. J. Gould, "The Archaeopteryx Flap," *Natural History*, p. 18, Sept. 1986.
26. Letter from Dr. Colin Patterson to Luther Sunderland, April 10, 1979.
27. Mark Ridley, "Who Doubts Evolution?" v. 90, *New Scientist*, p. 831, June 25, 1981.
28. Colin Patterson, Book Review, *Systematic Zoology*, v. 29 pp. 216-217, 1980.
29. Quoted in *The Neck of the Giraffe*, p. 16, 1983.
30. G. G. Simpson, *Life of the Past*, p. 119, 1953.
31. G. G. Simpson, *The Major Features of Evolution*, p. 263, 1953.
32. David Raup, "Conflicts Between Darwin and Paleontology," *Field Museum of Natural History Bull.*, v. 25, Jan. 1979.
33. Ibid.
34. E. Saiff and N. Macbeth, *Evolution* (unpublished ms.), 1982.

35. Francis Hitching, "Was Darwin Wrong?" *Life Magazine*,
    v. 5,  pp. 48-52, Nov. 4, April 1982.
36. Lyall Watson, "The Water People," *Science Digest*, v.
    90, May 1982.
37. Charles Darwin, *The Origin of Species*, 1859, p. 280,
    repr. 1966.
38. Cited in *Creation or Evolution*, Cliford Wilson, p. 68,
    1985.
39. Richard Leakey and Roger Lewin, *People of the Lake*,
    p. 26, 1978.

## CHAPTER TEN

1.  S. J. Gould, *The Panda's Thumb*, p. 188, 1982.
2.  S. J. Gould, "Evolution's Erratic Pace," *Natural History*,
    v. 5, p. 14, May 1977.
3.  Niles Eldredge and I. Tattersall, *The Myths of Human
    Evolution*  p. 46, 1982.
4.  Cited in *The Origin of Species Revisited*, p. 44, 1989.
5.  S. J. Gould, "Evolution's Erratic Pace," *Natural History*,
    v. 5, p. 12, May 1977.
6.  G. G. Simpson, *The Major Features of Evolution*, p.
    360, 1953.
7.  Ernst Mayr, (1970) *Populations, Species and
    Evolution*, p. 253. 1970.
8.  S. J. Gould, "The Five Kingdoms," *Natural History*,
    p. 37, June-July 1976.
9.  S. J. Gould, "The Return of Hopeful Mosters," *Natural
    History*, June-July, 1977.
10. Michael Denton, *Evolution: A Theory in Crisis*, p. 194,
    1986.

11. G. G. Simpson, *The Meaning of Evolution*, p. 231, 1949.
12. Stephen Stanley, *Macroevolution: Pattern and Process*, p. 35, 1979.
13. S. J. Gould, *The Panda's Thumb,* p. 186, 1982..
14. Ibid. p. 188.
15. Quoted in *What is Creation Science?*, Henry Morris and Gary Parker, p. 148, 1982.

## CHAPTER ELEVEN

1. Cited in *The Deluge Story in Stone*, Byron C. Nelson, p. 7, 1968.
2. Ibid., p. 9.
3. Ibid., p. 21.
4. John Whitcomb and Henry Morris, *The Genesis Flood*, pp. 67, 68, 1971.
5. John Whitcomb, *The World That Perished*, p. 22, 1988.
6. Henry Morris, et al., *A Symposium on Creation*, p. 110, 1968.
7. John Whitcomb and Henry Morris, *The Genesis Flood*, p. 21, 1971.
8. Ibid., Introduction
9. Byron Nelson, *The Deluge Story in Stone*, p. 98, 1968.
10. Carl Dunbar, *Historical Geology*, 2nd ed., p. 35-36, 39. 1961.
11. Cited in *The Deluge Story in Stone*, p. 42, 1968.
12. Ibid., p. 44
13. I. Velikovsky, *Earth in Upheaval*, p. 222, 1955.
14. N. O. Newell, "Adequacy of the Fossil Record," *Journal of Paleontology*, p. 492, May 1959.

15. Cited in *The Deluge Story in Stone*, p. 66, 1968.
16. Ibid., p. 85.
17. J. W. G. Johnson, *Evolution?* p. 67, 1986.
18. Duane Gish, *Evolution: The Challenge of the Fossil Record*, p. 50, 1985.
19. Heribert Nilsson, *Synthetische Artbildung*, pp. 1195-1196, 1953.
20. Kenneth Reese, "Newscripts," *Chemical and Engineering News* v. 11, p. 40, Oct. 1976.
21. Byron Nelson, *The Deluge Story in Stone*, p. 40, 1968.

## CHAPTER TWELVE

1.  Byron Nelson, *The Deluge Story in Stone*, p. 123, 1968.
2.  *New York World*, June 1, 1930.
3.  Jody Dillow, "The Catastrophic Deep-Freeze of the Beresovka Mammoth," *Creation Research Society Quarterly*, v.14,  pp. 5-12, June 1977.
4.  Ivan Sanderson, "Riddle of the Frozen Giants," *Saturday Evening Post,* p. 82, Jan. 16, 1960.
5.  Albert Sippert, *From Eternity to Eternity*, p. 107.
6.  Charles Hapgood, "The Mystery of the Frozen Mammoths," *Coronet*,  p. 74, Sept. 1960.
7.  John Whitcomb, *The World That Perished*, p. 80, 1988.
8.  Byron Nelson, *The Deluge Story in Stone*, p. 129, 1968.
9.  Ibid.
10. Ibid., p. 131.
11. Norman Macbeth, *Darwin Retried*, p. 115, 1971.
12. Dennis Petersen, *Unlocking The Mysteries of Creation*, p. 28, 1988.
13. Albert Sippert, *From Eternity to Eternity*, pp. 58-59,

1989.
14. Byron Nelson, *The Deluge Story in Stone*, p. 24, 1968.
15. Richard Carrington, *The Story of our Earth*, p. 163, 1956.
16. Byron Nelson, *The Deluge Story in Stone*, p. 148, 1968.
17. *Morrisonville Times*, June 11, 1891, Morrisonville, IL
18. Erich A. von Fange, "Time Upside Down," *Creation Research Society Quarterly*, p. 16, June 1974..
19. Duane Gish, *Acts and Facts*, v. 1, no. 4.
20. Paul Ackerman, *It's A Young World After All*, p. 106, 1988.

## CHAPTER THIRTEEN

1. Dave Balsiger and Charles Sellier, Jr., *In Search of Noah's Ark*, p. 212, 1976.
2. Ibid., p. 75.
3. Ibid.
4. St. Theophilus, *Ad Autolycum*, Book 3, Passage from M. Dods' trans. in the *Ante-Nicene Fathers*, p. 117, 1885, II.
5. From a sermon by John Chrysostom, Latin trans. from the Greek text by Migne, *Patrolograe cursus completus...series graeco*, LVI, cols. 287-88.
6. Isidore's *Etymologies*, Trans. from v. II of W. M. Lindsay's edition of the Latin text, 1911.
7. Marco Polo, *The Travels*, ed. and trans. Ronald Latham, p. 34, (The Folio Society, London, 1968).
8. Translated from Struys' account of his third voyage, Chapter 18 in *Les Voyages...en Moscovie, en Tartarie, en Perse, aux Indes, and en plusieurs*

*autres Pays etrangers,* pp. 146-62, 1684, II.
9. D. Webber, *Mystery of the Great Pyramid*, pp. 1-2.
10. Quoted in *The Deluge Story in Stone*, p. 165, 1968
11. D. Balsiger and C. Sellier, Jr., *In Search of Noah's Ark*, p. 70, 1976.
12. *Chicago Tribune*, Aug. 10, 1883, Chicago, IL.

## CHAPTER FOURTEEN

1. H. Brown, V. Monnett, and S. Stovoll, *Introduction to Geology*, p. 11, 1958.
2. Michael Pitman, *Adam and Evolution*, pp. 235, 236, 1987.
3. Henry Morris, *The Biblical Basis for Modern Science*, p. 310, 1984.
4. John Whitcomb and Henry Morris, *The Genesis Flood*, p. 187, 1971.
5. *Bible Science Newsletter*, July 1979.
6. Ibid.
7. M. K. Hubbert, and William Rubey, "Role of Fluid Pressure in Mechanics of Over-thrust Faulting," *Bulletin of Geological Society of America*, v. 70, pp. 115-166, Feb. 1959.
8. John Whitcomb and Henry Morris, *The Genesis Flood* p. 199, 1971.
9. Byron Nelson, *The Deluge Story in Stone*, p. 143, 1968.
10. *Bible Science Newsletter*, p. 6, Dec. 1986.
11. R. H. Rastall, "Geology," *Encyclopaedia Britannica*, v.10, p. 168, 1956.
12. Walter Brown, *In the Beginning*, p. 5, 1989.
13. David Milne, and Steven Schafersman, "Dinosaur

Tracks, Erosion Marks and Midnight Chisel Work (But no Human Footprints) in the Cretaceous Limestone of the Paluxy River Bed, Texas," *Journal of Geological Education*, v. 31, pp. 111-123, 1983.

14. Carl Baugh, *Dinosaur*, pp. 11-12, 1991.
15. W. A. Criswell, *Did Man Just Happen?*, p. 120, 1973.
16. Norm Sharbaugh, *Ammunition*, p. 183, 1991.
17. Ibid.
18. A. E. Wilder-Smith, *The Natural Sciences Know Nothing of Evolution*, p. 166, 1981.
19. Richard Leakey, Introduction, to *The Origin of Species*, R. Leakey ed., p. 15, 1979.
20. D. Axelrod, *Science*, v. 128, p. 7, 1958.
21. J. W. Klotz, *Genes, Genesis and Evolution*, pp. 193-194, 1972.
22. Henry Morris, *The Biblical Basis for Modern Science*, p. 339, 1984.
23. G. G. Simpson, *The Meaning of Evolution*, p. 18, 1949.
24. Ibid. p. 231.
25. Charles Darwin, *The Origin of Species*, p. 344. 1859.
26. Ibid., pp. 350-351.
27. Niles Eldredge, *The Monkey Business: A Scientist Looks at Creationism*, p. 44, 1982.
28. Ibid., p. 46.
29. H. Smith, *From Fish to Philosopher*, p. 26, 1953.
30. Personal letter to Luther Sunderland dated April 10, 1979.
31. David Kitts, *Paleobiology*, v. 5, pp. 353-354, Summer, 1979.
32. Niles Eldredge, *American Museum of Natural History*

*Guardian,* Nov. 21, 1979.
33. Charles Darwin, *The Origin of Species*, 1859 repr. 1958 New York American Library, p. 280.
34. W. H. Twenhofel, *Principles of Sedimentation*, 2nd ed., p. 144, 1950.
35. W. A. Tarr, "Meteorites in Sedimentary Rocks?" *Science*, v. 75, pp. 17-18, Jan. 1, 1932.
36. Walter Brown, *In The Beginning*, p. 53, 1989.
37. E. J. H. Corner in *Evolution in Contemporary Botanical Thought*, p. 61, 1961.

## CHAPTER FIFTEEN

1. Carl Dunbar, *Historical Geology*, 2nd ed., p. 18, 1960.
2. N. Geisler and J. K. Anderson, *Origin Science*, p. 106, 1987.
3. Merson Davies, "Scientific Discoveries and Their Bearing on the Biblical Account of the Noachian Deluge," *Journal of the Transactions of the Victorian Institute*, LXII, pp. 62-63, 1930.
4. George Wald, "The Origin of Life," *Scientific American*, p. 48, Aug. 1954.
5. John Ross, Letter, *Chemical and Engineering News*, p. 40, July 7, 1980.
6. R. Jastrow, *God and the Astronomers*, p. 111, 1978.
7. Arthur Eddington, *The Nature of the Physical World*, p. 130, 1930.
8. S. J. Gould, "Is Uniformitarianism Necessary?" *American Journal of Science,* v. 263, 226, March 1965.
9. S. J. Gould, "Catastrophes and Steady State Earth,"

*Natural History*, v. LXXX, No. 2, Feb. 1975.
10. James Shea, "Twelve Fallacies of Uniformitarianism"
     *Geology*, v.10, p. 456, Sept. 198.
11. Ibid., 457
12. Niles Eldredge, *The Monkey Business*, p. 96, 1982.
13. *National Geographic Magazine*, p. 128, July, 1962.
14. *National Geographic Magazine*, p. 715, May, 1972.
15. Steven Austin, "Mount St. Helens and Catastrophism,"
     *Impact*, July 1986.
16. *Reader's Digest*, Sept. 1977.
17. Ibid.


## CHAPTER SIXTEEN


1.  Anon., "What Future for Biogenesis?," v. 216 *Nature*
     p. 635, 1967.
2.  Letter from Charles Darwin to Sir Joseph Hooker
     (1871), in 2 *The Life and Letters of Charles Darwin*
     p. 202 (F. Darwin ed. 1887)..
3.  Charles Darwin, *The Origin of Species*, 1859 repr.
     1958, New American Library, p. 450.
4.  Quoted in *The Rise of the Evolution Fraud* by Bowden,
     p. 152, 1982.
5.  Yockey, "A Calculation of the Probability of
     Spontaneous Biogenesis by Information Theory," v.
     67 *J. Theoretical Biology,* p. 377, 1977.
6.  I. Prigogine, et al, "Thermodynamics of Evolution,"
     *Physics Today*, p. 23, Nov. 1972.
7.  J. Monod, *Chance and Necessity*, p.136, 1972.
8.  R. Jastrow, *Until the Sun Dies*, p. 60, 1977.
9.  Charles Thaxton, Walter Bradley and Roger Olsen, *The*

*Mystery of Life's Origin: Reassessing Current Thories*, Foreword by D. Kenyon, 1984.

10. Quoted in *Moody Monthly*, p. 21, Sept. 1988.
11. F. Hoyle, *Nature*, p. 105, Nov. 12, 1980.
12. F. Hoyle, "The Big Bang in Astronomy," v. 91 *New Scientist* p. 526, 1981.
13. D. E. Hull, "Thermodynamics and Kinetics of Spontaneous Generation," *Nature*, v. 186, p. 694, May 28, 1960.
14. R. Shapiro, *Origins: A Skeptic's Guide to the Creation of Life on Earth*, p. 112, 1986.
15. J. Brooks and G. Shaw, *Origins and Development of Living Systems*, p. 360, 1973.
16. Carl Sagan, v. 10, *Encyclopaedia Britannica: Macropaedia*, p. 894, 15th ed., 1974.
17. Haskins, "Advances and Challenges in Science in 1970," v.. 59, *American Scientist* p. 305, 1971.
18. F. Hoyle, *The Intelligent Universe*, p. 20-21, 1983.
19 Carl Sagan, "Life Beyond the Solar System," *Exobiology*, C. Ponnamperuma, ed., p. 471, 1972.
20. Quoted in *Evolution?*, p. 96, 1986..
21. Ibid.
22. *The Sunday Mail*, Brisbane, Aust., Sept. 20, 1981.
23. Ibid.
24. Quoted in *Evolution?*, p. 97.
25. G. G. Simpson, "The Nonprevalence of Humanoids," *Science*, CXLIII, p. 772, 1964.
26. F. Hoyle and N. Wickramasinghe, *Evolution from Space*, p. 148, 1981.
27. Quoted in *In the Beginning* by W. Brown, p. 41, 1989.
28. Quoted in *The Collapse of Evolution*, p. 3, 1988.

29. George Wald, "The Origin of Life," *The Physics and Chemistry of Life*, by the editors of *Scientific American*, p. 12, 1955.
30. George Wald, "The Origin of Life," *Scientific American*, p. 46, Aug. 1954.
31. Ibid.
32. George Marsden, "Creation verses Evolution, No Middle Way," *Nature*, v. 305, p. 572, Oct. 13, 1983.

## CHAPTER SEVENTEEN

1. W. Bird, *The Origin of Species Revisited*, p. 155, 1989.
2. S. Luria, S. Gould and S. Singer, *A View of Life*, p. 775 1981.
3. D. Raup, "Conflicts between Darwin and Paleontology," *Field Museum of Natural History Bull.*, p. 29, Jan. 1979.
4. Ibid.
5. Cited in *Darwin on Trial*, p. 20, 1991.
6. Cited in *Origin Science*, p. 84, 1987.
7. E. Russell, *The Diversity of Animals*, p.124, 1962.
8. Hughes and Lambert, "Functonalism, Structuralism, and Ways of Seeing," v. III *Journal of Theoretical Biology*, p. 796, 1984.
9. Luther Burbank, *Partner of Nature,* pp. 98-99, W. Hall ed., 1939.
10. W. R. Fix, *The Bone Peddlers: Selling Evolution,* pp. 184-185, 1984.
11. Deevey, The Reply: Letter from Birnam Wood, v. 61, p. 636, 1967.
12. P. Grasse, *The Evolution of Living Organisms*, p. 84,

trans. 1977.

13. L. H. Matthews, Intorduction to *The Origin of Species*, by Charles Darwin, J. M. Dent and Sons, Ltd., p. X, 1971.

14. M. Burton  and R. Burton, eds., *The International Wildlife Ency.*, p. 2706, 1970.

15. Quoted in *The Rise of the Evolution Fraud* by Bowden, p. 205, 1982.

16. Henry Morris, *The Twilight of Evolution*, p. 43, 1963..

17. P. Grasse, *The Evolution of Living Organisms*, p. 88 trans. 1977.

18. N. H. Nilsson, *Synthetische Artbildung*, p. 1157,  1953.

19. Ibid., p. 1186.

20. M. Pitman, *Adam and Evolution*, p. 67-68,  1984.

21. *2 The Life and Letters of Charles Darwin*, p. 210, F. Darwin ed., 1887.

22. Henry Morris, *The Twilight of Evolution*, p. 44, 1963.

23. J. J. Fried, *The Mystery of Heredity*,  pp. 135-136, 1971.

24. Cited in *The Neck of the Giraffe*, p. 64, 1983..

25. D. T. Rosevear, "Scientists Critical of Evolution," *Evolution Protest Movement Pamphlet,* no. 224, p. 4, July 1980.

26. Edward Blick, *Special Creation vs. Evolution*,  p. 38, 1981.

27. P. S. Moorhead and M. M. Kaplan, *Mathematical Challenges to the Neo-Darwinism Interpretation of Evolution*, Wistar Symposium no. 5,  1967.

28. Gordon Lindsay, *Evolution: The Incredible Hoax*, p. 16, 1977.

29. "At Random: A Television Preview," in *Issues in*

*Evolution,* v. III of *Evolution after Darwin,* Sol
Tax, ed.,.p. 41, 1960.
30. Colin Patterson, "Cladistics," BBC Interview, Mar. 4,
1982.

## CHAPTER EIGHTEEN

1.  Sir Solly Zuckerman, *Beyond the Ivory Tower*, p. 64,
1970.
2.  Charles Darwin, *The Descent of Man*, p. 213, 1871.
3.  G. G. Simpson, "The World into Which Darwin Led Us,"
*Science*, v. 131,  p. 969, 1969.
4.  Quoted by Michael Pitman in *Adam and Evolution*, p.
100, 1984.
5.  R. Leakey and R. Lewin, *Origins*, p. 106, 1977.
6.  David Pilbeam, "Rearranging Our Family Tree," *Human
Nature*, p. 45,  June 1978.
7.  "Famous Fossil Is a Fake Scientists Claim," *New York
Times*, reprinted in *Observer-Reporter*, Washington,
PA, May 9, 1985.
8.  Ibid.
9.  Quoted in *Neck of the Giraffe* by Hitching, p. 180,
1983.
10. Scott Huse, *The Collapse of Evolution*, p. 99, 1988.
11. A. Montagu, *Man: His First Million Years*,  2nd. ed., p.
58, 1962.
12. Cited in *Adam and Evolution*  by Pitman, p. 87, 1984.
13. E. A. Hooton, *Up from the Ape*, p. 329, 1946.
14. G. Himmelfarb, *Darwin and the Darwinian Revolution*,
p. 311, 1959.
15. W. R. Thompson, Introduction to *The Origin of*

*Species*, p. 17, 1956.
16. Duane Gish, *Evolution: The Challenge of the Fossil Record*, p. 188, 1986..
17. *Omaha World-Herald*, Feb. 24, 1928 Omaha, NE.
18. Ibid.
19. Solly Zuckerman, *Beyond The Ivory Tower*, pp. 75-94, 1970 and Oxnard, C. E., *Homo*, v. 30, p. 243, 1981.
20. Quoted in *Creation or Evolution* by C. Wilson, pp. 23-24, 1985.
21. H. Enoch, *Evolution or Creation*, p. 85, 1976,
22. Duane Gish, "Richard Leakey's Skull 1470," *ICR Impact*, no. 11.

## CHAPTER NINETEEN

1. Carl Dunbar, *Historical Geology*, p. 47, 1960.
2. Ibid.
3. R. H. Rastall, "Geology," *Encyclopaedia Britannica*, v. 10, p. 168, 1956.
4. *Scientific American*, v. 7, no. 38, p. 298, June 5, 1852.
5. Gary Parker, *Creation The Facts of Life*, p. 129, 1980.
6. Sigurdur Thorarinsson, *Surtsey*, p. 39, 1964.
7. Paul Ackerman, *It's A Young World After All*, p. 18, 1986.
8. Raymond Lyttleton, *The Modern Universe*, p. 72, 1956.
9. Isaac Asimov, "14 Million Tons of Dust Per Year," *Science Digest*, p. 36, Jan. 1959.
10. *The Rand McNally New Concise Atlas of the Universe*, p. 41, 1978.
11. Robert Dixon, *Dynamic Astronomy*, p. 149, 1971.
12. P. Stevenson, "Meteoritic Evidence for a Young Earth,"

*Creation Research Quarterly* 12,  June 1975, p. 23,
cited in R. L. Wysong, *The Creation-Evolution
Controversy*, p. 171, 1976.

13. Henry Faul, *Ages of Rocks, Planets, and Stars*, p. 61,
1966.

14. John Funkhouser and John Naughton, "Radiogenic
Helium and Argon in Ultromafic Inclusions from
Hawaii," *Journal for Geophysical Research* v. 73,
no.14,  pp. 4601-4607, July 15, 1968.

15. W. D. Stansfield, *The Science of Evolution*, p. 84, 1977.

16. Curt Teichert, "Some Biostratigraphical Concepts," *Bul.
of the Geological Soc. of Amer.*, v. 69,  p. 102, Jan.
1958.

17. Sylvia Baker, *Evolution: Bone of Contention*, p. 25,
1986.

18. Robert Gentry, "On the Invariance of the Decay Constant
Over Geological Time," *Creation Research Society
Quarterly*, v.. 5, pp. 83-84, Sept. 1968.

19. Cited in *Evolution?* by Walter Johnson,  p. 76, 1986.

20. Dennis Petersen, *Unlocking The Mysteries of Creation*,
p. 20, 1988.

21. C. A. Reed, "Animal Domestication in the Prehistoric
Near East," in *Science*, v. 130, p. 1630, 1959.

22. Caryle P. Haskins, "Advances and Challenges in Science
in 1970," *American Scientist*, v. 59, p. 298, May-
June 1971.

23.  Dennis Petersen, *Unlocking The Mysteries of Creation*,
p. 49, 1988.

24. Ibid.

25. Robert E. Lee, "Radiocarbon Ages in Error," *Anthropol-
ogical Journal of Canada*, v. 19, no. 3, p. 9, 1981.

26. Ibid. p. 29.
27. Ariel Roth, "Coral Reef Growth," *Origins*, v. 6, no. 2, pp. 88-95, 1979.
28. John Whitcomb, *The World That Perished*, p. 114, 1973.
29. Ibid., p. 135.
30. R. Wysong, *The Creation-Evolution Controversy*, p. 455, 1984.
31. Edward Blick, *Correlation of the Bible and Science*, p. 28, 1976..
32. Donald Patten, *The Biblical Flood and The Ice Epoch*, p. 11, 1966.
33. Dennis Petersen, *Unlocking The Mysteries of Creation*, p. 43, 1988.
34. Henry Morris, W. W. Boardman and R. Koontz, *Science and Creation*, p. 75, 1971.
35. Kent Hovind, videotape #1, Creation lecture, 1992.

## CHAPTER TWENTY

1. *Evolution?*, p. 26, 1986.
2. Henry Morris, *The Biblical Basis for Modern Science*, p. 353, 1984.
3. G. G. Simpson, quoted in *Life Before Man*, p. 42, 1972.
4. Cited in *Unlocking The Mysteries of Creation*, by D. Petersen, p. 149, 1988.
5. *Minneapolis Tribune*, Jan. 6, 1982, Minn. MN.
6. David Milne and Steven Schafe

*of Geological Education* v. 31, pp. 111-123, 1983.
7.  Carl Baugh, *Dinosaur,* p. 11, 12, 1991.
8.  *The Somerwell Sun*, June 24, 1987, Somerwell Co., TX.
9.  Y. Kruzhilin and V. Ovcharov, "A Horse from the Dinosaur Epoch?" *Moskovskaya Pravda* (Moscow Truth), Feb. 5, 1984.
10. V. Rubston, "Tracking Dinosaurs," *Moscow News*, no. 24, p. 10, 1983.
11. Melvin Cook, "William J. Meister Discovery of Human Footprints with Trilobites in a Cambrian Formation of Western Utah," in *Why Not Creation?* ed. Walter Lammerts, pp. 185-193, 1970.
12. "Geology and Ethnology Disagree About Rock Prints," *Science News Letter*, p. 372, Dec. 10, 1938.
13. Henry Schoolcraft and Thomas Benton, "Remarks on the Prints of Human Feet, Observed in the Secondary Limestone of the Mississippi Valley," *American Journal of Arts and Sciences*, v. 5, pp. 223-231, 1822.

## CHAPTER TWENTY-ONE

1.  Henry Morris, *A History of Modern Creationism*, p. 263, 1984.
2.  Niles Eldredge, *The Monkey Business*, p. 17, 1982.
3.  S. J. Gould, *Natural History*, p. 19, April 1993.
4.  H. L. Mencken, *Prejudices: Fifth Series*, (New York, 1926), from selection in Henry May, ed., *The Discontent of the Intellectuals: A Problem of the Twenties*, p. 29, 1963.
5.  A. Rehwinkel, *The Wonders of Creation*, p. 31, 1974.

6.  *Acts and Facts*, p. 2, August 1982.
7.  A. T. J. Hayward, *Nature*, Correspondence,  v. 240, p. 557,  Dec. 29, 1972.
8.  Henry Morris, *The Battle for Creation*, p. 83, 1976.
9.  Clifford Burdick, "Microflora of the Grand Canyon," *Creation Research Society Quarterly*, v. 3, no. 1, pp. 38-50,  May 1966.
10. Robert Gentry, *Creation's Tiny Mystery*, Foreword  by Dr. W. Scot Morrow, 1986.
11. Frank T. Manheim, "Letter," *Geotimes,* p. 4, April, 1993.
12. J. Lane, "Letters: Creationism discussion continues," *Physics Today*, p. 103, Oct. 8, 1982.
13. Charles Darwin, *The Origin of Species*,  p. 482 1st ed. 1859, repr. 1964.
14. R. Bozarth, "The Meaning of Evolution," *The American Atheist,*  v. 20, no. 2, p. 30, Feb. 1978.

## EPILOGUE

1.  NBC News and Associated Press November National Poll 15, Nov. 24, 1981.
2.  Cited in *Moody Monthly*, p. 20, Sept. 1988.
3.  *The World's Most Famous Court Trial*,  p.187, 3rd ed. 1925, (trans. of Scopes Trial).
4.  P. Davis and E. Solomon, *The World of Biology*, p. 414 1974.

# NAME AND SUBJECT INDEX

Books by the Author:

**Liberalism: A Rope of Sand**, 203 pages, $8.00--gay rights, welfare, death penalty, pornography, public schools, sex ed., etc.

**Your Health: How to Feel Better, Look Younger, Live Longer**, 145 pages, $7.00--general health, good food, exercise, addiction, family health, mental health, etc.

**Pilgrims, Puritans, & Patriots**, 247 pages, $8.00--colonial history, what Pilgrims and Puritans believed, Great Awakening, slavery, training of Puritan children, etc.

**Is God a Right-Winger?**, 268 pages, $8.00,--welfare, taxes, perversion, criminals and crime, Martin Luther King, Jr., Senator Joe McCarthy, Christian schools, abortion, etc.

**Christian Resistance: An Idea Whose Time Has Come--Again!**, 81 pages, $4.00--What does Romans 13 teach?, What did the early Church teach and practice?, Should guns ever be used?, etc.

**Pioneers, Preachers and Politicians**, 402 pages, $14.00--a full American history highlighting contributions made by Christians and Bible preachers.

**Boys' Big Book of Humor**, 221 pages, $8.00--a delightful book of one liners, tombstone humor, limericks, education humor, religious humor, love and marriage humor, etc.

**AIDS: Silent Killer**, 256 pages, $10.00--the AIDS coverup, homosexuals and AIDS, the cost, what should we do, AIDS policy for churches, schools, etc.

**Evolution: Fact, Fraud, or Faith?**, 352 pages, $15.00--book exposing evolution as a fake, a fraud, a faith and a lot of foolishness.

Send check or money order to Freedom Publications, P. O. Box 981, Largo, FL 34649. Prices subject to change without notice. Add 15% ($1.50 minimum) for shipping and handling.